IEE PROFESSIONAL APPLICATIONS OF COMPUTING SERIES 2

Series Editors: Professor P. Thomas
Dr. R. Macredie

UML for systems engineering

watching the wheels

UML for systems engineering

engineering

watching the wheels

Jon Holt

The Institution of Electrical Engineers

Published by: The Institution of Electrical Engineers, London,
United Kingdom

British Library Cataloguing in Publication Data

Holt, John
UML for systems engineering: watching the wheels
(IEE professional applications of computing series; no. 2)
1. UML (Computer science) 2. Systems engineering
I. Title II. Institution of Electrical Engineers
005.1' 17

ISBN 0 85296 105 7

Printed in England by
Antony Rowe Ltd, Chippenham, Wiltshire

This book is dedicated to my mother,
Christine Holt

Contents

PART II. Talking the language

Chapter 2   Modelling

Foreword

Throughout the ages, engineers and technologists of all disciplines have been responsible for the development of various systems of immense complexity and deep social significance. Unmanned space probes, nuclear-powered submarines, the mind-blowing Mayan temples and the pyramids of Egypt, dams that enclose waters that are larger than many natural seas, with their integrated hydroelectric generation and distribution systems, and heart-transplant life-support equipment all bear witness to human creativity and the ability to marshal multitudes of components to form massively complicated functional systems.

Provocatively, one could suggest that there *is* actually no such discipline as "Systems Engineering"—for "Engineering" is inherently always about the development of systems!

The 21st Century continues this long history of technological innovation and human achievement. However, in an increasingly open and competitive world, resulting from dramatically increasing global connectivity between individuals, companies and governments, there is ever-increasing pressure to minimise the whole engineering cycle—from initial idea through to fully functional and maintained product. At the same time, though, global competition forces engineers to ensure that the system being produced is right-first-time, that it is delivered on time and within budget, and that it has guaranteed and predictable performance.

It therefore comes as no surprise that designers of all sorts of complex systems are continually seeking tools to support each aspect of their work—starting with gaining a proper understanding of what is required of the system to be developed, right through to final product validation and proof of conformance—both to the initial requirements and also to any other applicable legal or technological requirements.

Initially emerging from efforts to produce tools for the life-cycle support of software, where the levels of complexity (and the day-to-day experience of our inability to master it!) are very evident, has come the Unified Modelling Language, UML. This powerful set of modelling tools and techniques has gained rapid acceptance in the software engineering community. That, in itself, is interesting, given the healthy scepticism with which the practitioners have greeted most other such tools! This acceptance is probably a result of the inherent simplicity and intuitiveness of the various UML techniques, combined with the fact that, together, they can be used in a highly integrated fashion, allowing various contributors or stakeholders in any project to "view" the proposed system from various different points of view.

In this unique book, Jon Holt draws on his wide practical experience in applying both the UML and many of its predecessors (especially OMT), initially in software-based systems, to take the reader into a much broader world than those suggested by most proponents of the UML—by applying the language to systems design in general. Through practical examples, he shows how the UML can be applied to a range of non-software-based applications. (Of course, many of the systems that can be specified, analysed and designed using the UML may and probably will have software components, but this will be determined by the UML-based design work, and need not be an assumed implementation requirement.)

A particularly intriguing area of the book, which arises when exploring the broader use of the UML, deals with the description and analysis of technical standards and their associated documentation. Jon Holt's work in this area has considerable potential, especially in an era when technology is advancing so fast that the development of supporting standards often lags well behind the practical possibilities. The result, as is pointed out in this work, is often the publication of standards that are incomplete, inconsistent or, simply, wrong! The use of the UML can go a long way towards ensuring far greater transparency and understanding of standards, and will certainly make them far simpler to analyse.

Whilst many excellent introductions to the UML are now available, this present text provides a unique, entertaining and highly readable insight into the range of instances where the language might feasibly be used, and how it can underpin the activities of developers of all complex systems. It moves the UML out of the world of software into the wide world of "Engineering" in its broadest interpretation—including that of (so-called) "Systems Engineering"!

Mike Rodd, PhD, DSc
Director of Knowledge Services, IEE
and
Honorary Professor, University of Wales, Swansea

Preface

This book is the result of many years' work and research into modelling and the UML. The book came about because of the UML courses that I have developed and taught over the last two years and the feedback from the several hundred people that have attended them. So many of these people have expressed the need for a full book based on systems applications using the UML, so here it is.

The book itself is not intended to replace any existing UML texts, but is intended to be used in conjunction with them. The main aim, however, is to try to bring the UML to a wider audience (systems engineers) than its traditional software-based following. The intended audience for this book is, therefore, practising systems engineers, systems engineering students, systems engineering managers and anyone who is interested in making wider use of the UML.

As a result of this systems emphasis, there are no in-depth software applications in the book. Neither are there any great comparisons with other modelling techniques, and this is for two reasons: many other books do this so there is no point reinventing the wheel; the UML is an industry standard that is being used by a continuing audience of people who need to communicate effectively, and therefore it cannot be ignored.

The UML presented here represents only about 10–15 per cent of the total syntax, but this is intentional. My advice is to learn a small amount of UML and to understand it! Next, use some of the other texts that are available to fill the gaps in your knowledge.

From a systems engineering point of view, the book is not intended to be a definitive systems engineering book, rather a refresher of some important systems concepts that will affect all systems engineers. The systems concepts that are covered in this book are by no means complete, but, based on current areas of interest from organisations such as the IEE and INCOSE, it tries to tackle some of the main issues.

The book is called '*watching the wheels*' for two reasons. The first is because of the John Lennon song that contains the line "there's no problems, only solutions", which is inspirational for systems engineers everywhere. The second is slightly more esoteric, as wheels do not occur in nature. Wheels represent man-made complexity, especially when they start to interact.

As with anything in life, there is always room for improvement, so if you have any comments, criticisms or observations, please feel free to contact me. Above all, enjoy the book!

Jon Holt
Brass Bullet Ltd
(jon@brass-bullet.co.uk)

Acknowledgements

No work is carried out in isolation and this book is no exception. If I were to acknowledge everyone who had contributed to this book in some way (either directly or indirectly) then I would have to include everyone I have met over the last ten years! I shall make an attempt to mention as many people as possible, but if you know me and are not on the list, please accept my apologies.

First and foremost, I must thank my Brass Bullet colleagues: Guy Tugwell, Keith Ellsmore, Mike Brownsword and Rebecca Saltmarsh. I would also like to thank the IEE and, in particular, Mike and Sue Rodd (who taught me English and who still, after all these years, seem to have a little faith in me), Martin, Nancy and Michelle. While on the subject of the IEE, I would like to acknowledge the support and help of the now-defunct PG A6, especially Steve Willmott and the new PeNSE group. INCOSE deserves a mention as it has been instrumental in allowing me to promote the UML for systems engineering through its many excellent events and thus has my thanks.

I would like to thank all of our clients who shall not (or cannot!) be named, but who know who they are. However, one or two deserve to be singled out: extra thanks should go to Dipesh Patel and Mick McManus (no, not the wrestler) who have given me a free rein on their process for the last year or so; also, special thanks must go to Ian Wilson from Newcastle who has supported the courses, allowed us to try out new ideas and even proofread some of the chapters.

From a UML point of view, I would like to thank Grady Booch, James Rumbaugh and Ivar Jacobson, none of whom I have met, but all of whom have had a big influence on my work. Cheers guys!

From an academic point of view, many thanks go the Department of Electrical Engineering at the University of Wales Swansea, headed by the curry-fuelled Bob Cryan and ably assisted by Chris Jobling and John Mason.

Many thanks must also go to the IEE publishing department for all their help, 'discussion' and guidance, and their marvellous guidelines for preparing manuscripts!

Penultimate thanks go to Paul Labbett who provided the cover art and who will expect to be bought a drink for every copy of the book sold!

Finally, my undying love and thanks go to my family: Rebecca, Jude and baby Eliza, who continue to motivate and inspire my every action in life.

Jon Holt, July 2001
Brass Bullet Ltd
(jon@brass-bullet.co.uk)

List of figures

Chapter 4   Behavioural modelling

Chapter 5   The UML diagrams

Part III. Solutions

Chapter 6   Modelling standards, processes and procedures

Chapter 8   Extending the UML

PART I: Problems

Chapter 1

Systems engineering

"if you're looking for trouble, you came to the right place"
– Elvis Presley

1.1 Problems with systems

1.1.1 Defining systems engineering – first attempt

This book is concerned, primarily, with applying the Unified Modelling Language (UML) to systems engineering. It would seem appropriate, therefore, to give some definitions for systems engineering at this initial stage. However, this is not as simple as it may appear as there are many definitions to choose from and it would be precipitous to state a definition at this stage without giving some thought to the issues involved. It is not only important to understand the needs for systems engineering, but also to understand what a system is. Therefore, deriving a definition for systems engineering will be left until later in this chapter when we have more of an idea about the scope and complexity of the field of systems engineering. Rather than a single, definitive statement of what exactly systems engineering is, this book will put forward two definitions – one technical, one practical – which should always be kept at the back of your mind while going about life, but more of this later.

1.1.2 The need for systems engineering

It was stated in the introduction to this chapter that an understanding of why we need systems is required, together with an understanding of what, exactly, a system is [1, 2]. This section aims to address the first of these two points, by asking the question, why do we need systems engineering? Is there a real need or is it just some academic exercise to appear knowledgeable and to sell books?

In order to understand the need for systems engineering, let us first look at a number of the driving forces behind it. These include the following:

- Increased size of the global market and the global village. As the world is getting smaller and many boundaries, whether social, political or economical are removed, so the need to compete in this freer world becomes more important. The vast increase in the efficiency of communications over the last few years has meant that it is now possible to trade on a worldwide scale at the touch of a button. The presence of the Internet and, more to the point, its increased use by all aspects of society, means that if a product is cheaper in one country than another, customers do not have to leave the comfort of their armchairs to effectively go there and buy it. With virtual working it is now feasible to work 'closely' with people on the other side of the world, rather than being restricted to working with 'local' people. With such an increase in competition, it is important that the systems that we produce are the best that anyone anywhere can offer.

- Increased technological performance. As the speed and efficiency of technology increases, so the demand for such systems also increases. Customers will always want products that are smaller, more efficient and smarter than previous models and will accept no arguments as to why this is not so. It is crucial that these rapidly changing, advanced systems are managed and understood if they are to provide the safety and reliability required of a modern-day system.

- Ubiquitous computers. Computers are now an integral part of almost any complex system and will generally involve software of some description which is, arguably, the most complex of man's creations. Software must, therefore, be fully understood and managed as part of a large, complex system otherwise we run the risk of another software crisis. The world of software engineering has recently undergone a watershed of understanding, which has led to the realisation that the Mickey Mouse sorcerer's apprentice approach to writing code was simply not good enough. Software is an unpredictable beast that needs to be contained by applying software engineering techniques that increase the chance of project success enormously. However, when this software is released into an integrated complex system, software engineering techniques alone are not sufficient for the increased problem space, hence the variety and range of techniques that can control the beast must be increased accordingly. Software engineering is a subset of systems engineering that limits itself to a single discipline. Clearly, there are many similarities between the two disciplines and it is important to learn lessons from any source available.

- Quality. This must be an integral part of the whole life cycle of system development and will ensure the future of systems and their associated business [3]. One of the problems with quality is that it is often perceived as simply a 'certificate on a wall' or a 'badge with a profile drawn on it'. However, quality should be viewed as a frame of mind that should be adopted by all project personnel. Only when we think in terms of quality can we hope to produce systems that have quality that is in-built, rather than tacked on as an afterthought.

- Integration of systems. This brings together all the points listed above. Systems engineering encompasses all areas of engineering and management and, as such, requires new levels of understanding. This introduces problems of complexity, communication between different disciplines and is ripe for potential misunderstandings between project personnel. All of this, naturally, should be avoided.

- Finally, it must be remembered that massive losses are made each year due to project failure and disasters, many of which could have been avoided with the application of adequate systems engineering.

The points raised here are not complete but provide a brief introduction as to why people are sometimes caught unawares and can end up with a potential disaster on their hands once it is too late to do anything about it.

1.1.3   Disasters and failures

Left unchecked in any system, problems may lead to either failure or disaster. In order to understand this, consider the following two broad groups that categorise most results of problems:

- System disasters. By which we mean that something going wrong results in the loss of human life, damage to the environment or damage to property. In many cases, these are systems that have run successfully and error-free for many years, sometimes decades, before disaster strikes.

- Project failures. By which we mean a project that is either never completed or one that was abandoned soon after its initial operation.

Although there are many reasons behind these two types of system problem, there are a number of common themes that emerge from almost all examples of them, and these will be explored in the next section.

For examples of failures and disasters, see [4], [5], [6] and [7].

1.1.4   Causes of problems

History has taught us that, whatever the nature of disaster or failure, there are a few common causes that crop up time and time again. This conclusion has been reached by many authors and may be summarised by considering three points – complexity, communications and a lack of understanding – the 'three evils of systems engineering'.

- Complexity. This is certainly one of the main causes of system problems. This may be because the complexity of a system may have been underestimated or not taken into account at all [8]. It may also be that the complexity has been 'unmanaged', by which we mean that it may have been uncontrolled or unmeasured. In order to manage such complexity, however, it is important to have a good idea of what, exactly, complexity is – which will be discussed in some detail in a subsequent section. Mismanaging complexity will lead to a lack of understanding and bad communications.

- Communications. A lack of communication or inefficient communication will contribute to project failure. It is important to identify the communication needs in a project and then to communicate effectively. Poor communications will lead to ambiguities, which may lead to complexity. They will also lead to a lack of understanding, which is obvious if people cannot understand what one another are saying. Communications will be discussed in more detail later in this chapter.

- Lack of understanding. This is a problem that may occur for any number of reasons, such as a lack of experience, not understanding the domain, not speaking the right language, underestimating complexity, and so forth.

One point that should stand out after reading this list is that each of the problems is likely to lead to difficulties in the other two areas, or to put it another way, it is a circular problem. This means that we must try to address all of these issues in some way and keep the problems to a minimum. Half of the battle with these problems is trying to highlight where they occur during a project. At what stage then should we concentrate on trying to spot these areas and are they more likely to occur early on during the development of a project or in the later stages?

These problems do not simply occur during the design and implementation of a project, but may appear at any phase in the project's life cycle. The problems themselves are just as likely to happen at any point of project development, but the later they do, the more they become compounded. For example, by the design phase it is far more expensive to correct a fault than if it were found during the requirements phase. Therefore, the way in which we approach systems engineering must be applicable at any point in the project life cycle rather than simply focusing on a single phase or phase activity. This will turn out to be a fundamental requirement for systems engineering.

1.1.5   Defining a system

The motivation behind systems engineering has been discussed, but what do we actually mean by a system? A common misconception is that a system is simply a product that may be delivered at the end of a project. However, this is not the case. A system may include any or all of the following points:

- A product or set of products that are delivered to the customer as part of the overall system. The product may be a physical system that may be picked up and touched or it may be a virtual system such as software. A system may not necessarily be a technical system, but may be an ecosystem, a social system, a political system or an economical system. Taking matters to an extreme point, a system may be a human being, which is, arguably, one of the most complex systems in the world.

- Operational procedures or knowledge concerning the final product that will enable the customer to successfully install and run the system, and the end-users to successfully operate the system. It is important that this information is built into the system design, rather than being added after the system has been fully developed.

- Support and maintenance of the system, to take into account future refinements, error corrections, updates and servicing.

- Training to enable customers and end-users to use and operate the system in a safe and reliable fashion. This will not only ensure the safety and comfort of end-users, but may also prolong the life of the system itself.

- Integration with other systems, as it is crucial to the operation of the system to make sure that it fits seamlessly with other components in the operational environment.

A system, therefore, is more than a final product. It is a living, evolving entity that requires looking after and occasional help and therapy. In order to keep a system in top condition, a sound understanding of the way it both looks and behaves is required both under normal conditions and under atypical conditions.

1.1.6 Defining systems engineering – second attempt

Based on what has been discussed so far in this chapter, two definitions for systems engineering are suggested. The first definition of systems engineering that is derived directly from these discussions, taking all the points that were raised here, is:

"Systems engineering is concerned with defining and implementing an approach to solving problems, while managing complexity and communicating effectively over the entire lifetime of a project."

Another definition of systems engineering that has been derived indirectly from all the points raised in this chapter is:

"Systems engineering is the implementation of common sense."

This may seem like a strange definition, but all of the points that have been raised in this section are sound, basic engineering principles. In addition, when problems are analysed and the resultant causes identified, the initial reactions from most people are shock, horror and amusement that people could be so stupid to allow it to happen in the first place!

1.1.7 Summary

In summary, therefore, the following points have been raised so far in this chapter.

- There is a need to engineer systems in order to make them as safe and reliable as possible and in order to give people confidence in the systems that we produce.

- Many problems with systems have three common themes: complexity, communications and a lack of understanding. All three of the problem areas must be borne in mind when engineering a system.

- A system was defined not simply as a product, but as products with a set of associated services.

- The nature of a system need not be physical, as it could be an intangible system, such as software or a social system.

- Systems engineering was defined in the context of this book as an approach to solving problems, while managing complexity and communicating effectively over a whole life cycle.

- Systems engineering was also defined as the implementation of common sense.

This section has introduced the main issues involved with systems engineering and the following sections look at some of these issues in more detail. These issues are complexity, communications and understanding in Section 1.2, and quality in Section 1.3.

1.2 Complexity, communication and understanding

1.2.1 Introduction

This section looks at the points that were raised in the previous sections as major causes of problems and discusses them in more detail. The main proposal here is that if we can understand a little about the nature of these problems, hopefully we can go some way towards addressing them so that we do not allow history to repeat itself with yet more failures and disasters. The main aim is to try to draw up a set of requirements that, when met, will allow us to go some way towards solving these problems. The term 'some way' is used here as it is impossible to eliminate all problems, but we must still do as much as we can. This is the same theory as the one concerning testing a system, where it is impossible to test a system 100 per cent, it is only possible to test a system to give a degree of confidence that the system will work under all anticipated circumstances.

The first of the points to be addressed is that of complexity, which will be discussed in the following section.

1.2.2 *Complexity*

Complexity is currently a hot topic in systems engineering and there are many excellent texts available that describe it in great detail. This section aims to define complexity at a high level so that it can be understood, to a limited degree, by everyone reading this book. For more in-depth discussions and philosophical musing, readers are directed towards the titles referenced at the end of this chapter.

To start the discussion, it is necessary to visit the world of software engineering and to consider the work of Fred Brooks. Brooks, often viewed as the father of software engineering and one of the great software philosophers (if such a thing exists), identifies two main types of complexity that exist in almost all systems (software or otherwise): essential complexity and accidental complexity [9].

It is important to be able to distinguish between these two types of complexity, so that they can be identified, or potential areas identified, within a system.

1.2.2.1 *Essential complexity*

Essential complexity is so called, not because it is vital to the system, but because it is in the essence of the system. This means that it is an inherent part of a system and, as such, cannot be eliminated. Clearly, the presence of such complexity is a big problem, which means that it is even more important that these areas of essential complexity are identified and have attention drawn to them in order that their effect on the rest of the system be minimised.

As an example of essential complexity, consider any system using a PC or similar computer. Within a PC there is a level of complexity that will exist within the PC itself. There is nothing that can be done to minimise the complexity of the hardware of the PC, short of taking it apart and completely redesigning it. The next level of complexity that exists is that of the operating system where, again, it is not possible to do anything about, short of re-installing another operating system. It may even be the case that there is more than one operating system, although more recently developed operating systems now claim seamless integration with the hardware, rather than using a middle-level operating system. If someone wants to develop software on the PC, there is the complexity of an interactive development environment that enables the user to write and manage the code, together with the complexity of the language and the compiler itself. This gives a number of factors that cannot be changed in any way and, hence, such a system will exhibit essential complexity as it lies within the essence of the system. All this before even a single line of code has been written!

Another example may be a legacy system that may have to exist as part of a new system. In real life, systems tend to evolve rather than be started from scratch, and thus legacy systems are often a necessary evil. It is often impossible or unfeasible to change a legacy system in any way; therefore, essential complexity will always exist within the legacy system.

One argument that is often put forward is that if there is nothing that can be done about essential complexity, then what is the point of thinking about it in the first place? The answer to this concerns project risk. If all areas of complexity are identified, it is possible to perform a risk assessment exercise to determine any parts of the system that are critical to the success of the project. If these areas of complexity are known, it is possible to design the system so as to minimise the risk of the essentially complex parts.

1.2.2.2   *Accidental complexity*

Accidental complexity differs from essential complexity as, unlike essential complexity, which is inherently part of a system, accidental complexity is complexity that creeps into the system by accident. As accidental complexity is caused by accident or, to put it another way, by error (or even omission), it is possible to do something about it. In an ideal world, the solution would be to eliminate accidental complexity altogether, but in reality the best that can be hoped for is to minimise accidental complexity as much as possible.

In order to be able to minimise accidental complexity, it is crucial to understand how it may creep into a system. A number of causes of accidental complexity are discussed in the list below:

- Poor front-end life cycle activities. This includes, for example, requirements engineering, which drives the rest of the project. If the requirements are not right or have not been engineered adequately, complexity may creep in right at the beginning of the system's life cycle. This may be due to the fact that a requirement has not been analysed sufficiently and the complexity has been underestimated.

- A lack of understanding of the system. This is particularly prevalent in the problem domain when the analysis is being carried out, which will lead to accidental complexity. Many projects fail due to the fact that the problem is not understood well enough before the design is started, which means that the likelihood of creating a correct solution is very small indeed.

- A lack of effective communications. This may well lead to complexity and sets off a vicious circle as communications and complexity tend to chase each other's tails and each result in the other. Poor communications, particularly at earlier life cycle phases such as requirements and analysis may lead to all sorts of problems later on in the project. Even very simple terms will have meaning to some people but not others, or may have totally different meanings altogether.

Accidental complexity, therefore, is something that we can go a long way towards addressing and, as such, it is important to be able to minimise such complexity.

1.2.3 Minimising complexity

In an ideal world, complexity would be eliminated altogether. However, in reality this is almost always not the case and thus complexity must be first identified and then minimised.

Complexity may be minimised by effective communication and by applying that great maxim that impacts upon all areas of life, KISS, or 'keep it simple stupid!'

In order to do this, a well-defined approach is essential. This approach must be simple, straightforward and correct. By 'correct', we mean that the approach must be consistent within itself and comply with the way in which people work. The approach must also be known by all staff and interpreted by each team member in the same way.

1.2.4 Communications

1.2.4.1 Introduction

The subject of communications has already reared its ugly head at several points during this chapter, but what exactly is meant by communications? It is probably safe to say that everyone realises that communication is important for any project or any system, but what channels of communication exist? They exist at different levels and it is important to have an idea of what these levels are. These include:

- A channel of communication that exists at an organisational level. Many projects will consist of workers from more than one organisation, which has the potential to lead to communications problems. The languages used by organisations may differ, with both technical and spoken languages and the protocols involved with talking to other people or groups of people possibly differing radically between organisations.

- A channel of communication that exists between different teams within a single organisation. People may work in disparate areas of the company and will thus have a completely different vocabulary due to their different areas of technical expertise. In addition, communications between teams may be semi-deliberate for fear of competition from rival groups within a single organisation, for example. This may, in extreme cases, even lead to sabotage between groups as one group is desperate to keep up with another. This may take on a relatively harmless form of sabotage, such as withholding information, or a far more malicious form of sabotage, such as feeding deliberate misinformation to another team. A further aspect of communication between teams which may lead to problems is that people assume that other teams know something or have done something that may be crucial to the project. Making assumptions can be dangerous when it is taken for granted that everybody knows what they are, and they should, therefore, be explicitly defined and communicated.

- At a lower level of communication, a channel exists between different resources, such as hardware, software, networks, protocols, etc. This may manifest itself by a particular piece of hardware not working with other hardware or new versions of software not working with older operating systems or software. It may also be caused by legacy systems that have existed within a system for many years yet, for whatever reason, remain a crucial part of a system.

- Problems may also exist between different levels of management. Many reasons exist for this, such as fear of 'shooting the messenger'. People are often scared of reporting failure, even if it has nothing to do with them, as it may be perceived as being the direct fault of the messenger. This may also manifest itself because there are too many layers of management within an organisation, making effective communication impossible due to the complexity of getting a single message to a high level of management via several other managers. People will often gloss over problems, which may result in the snowball effect of a simple problem the size of a small snowball gently rolling down a hill for as long as it is ignored, before finally reaching the bottom of the hill where it can no longer be ignored, is the size of a house and destroys everything in its path or, in this case, leads to the demise of the project.

These levels of communication have been seen to have their own dangers within a project if not properly identified. It is crucial, therefore, to identify communication interfaces between the roles of people involved within a project. For example, who reports to whom within the organisation? What are the interfaces to other organisations and who is allowed to communicate across these boundaries? How are problems reported and to whom? Is blame associated with the reporting of problems or is the cause of the problem the main concern?

These difficulties can all be amplified by a single aspect of communication – that of the language being used.

1.2.4.2 Languages

One of the basic requirements for communicating between one or more entities is to have some sort of language that each entity can speak. Any problems with a language will compound the problems related in the previous section. Many problems with languages exist, including:

- People from different countries may speak different languages, hence making communication almost impossible without an interpreter. Of course, the actual interpreter may introduce a number of ambiguities into the translation, thus introducing additional complications and ambiguities.

- People with different working backgrounds may understand different things from the same term. For example, the word 'fencing' may apply to making fences and barriers, the noble art of swordplay or, indeed, selling stolen goods.

The meaning of each of these is completely different and such a misunderstanding could potentially lead to large problems within a project.

- Perhaps one of the most unpredictable problems occurs when people think that they speak the same language but, in reality, there are big differences in the meaning of words. An example of this is American-English vis-à-vis English. These two languages are deceptively similar but each has many words that either have no meaning in the other language or a completely different meaning. There is an old adage that the United States and the United Kingdom are separated by a common language and this is absolutely true. Many terms are taken for granted as having the same meaning when they do not. For example, think of a motor car that in the United Kingdom would have a bonnet and a boot, while in the United States an automobile would have a hood and a trunk.

The previous three points refer to spoken languages, but these are not the only types of language that may be used on a project, and problems may also exist as a result of differences in technical languages such as:

- Programming languages. In the case of software, there may be different versions, or flavours, of languages. Anyone who has ever used the C programming language, for example, will know that there are numerous definitions of the language itself, which will depend on many things, including the manufacturer chosen. Even when a language is chosen from a particular manufacturer, there will be different versions of the language, the compiler and the interactive development environment (where appropriate).

- Modelling languages. It was very much the case that before the advent of the UML, there were over 50 modelling languages, making inter-company communication extremely difficult. Even with a standard language, such as the UML, different people have different slants on the same concepts. Luckily, there is a single source of reference for the UML, so any ambiguities can be quickly sorted.

Clearly, there is a need for an unambiguous language that all interested parties can understand, which points towards a requirement for a common language.

1.2.4.3 Common languages

As the points raised in the previous section have made clear, there is a need for a common language. This relates to the old adage about 'everybody singing from the same sheet' where it is crucial that everyone understands the same thing by every term used. This also relates back to the famous 'Tower of Babel' story from the Bible (suggested as a good example by Brookes), which involved a group of people who wanted to build a tower up to heaven, so that they could talk to God (this is a somewhat abridged version of the Bible story!). According to the original story, the builders worked very well together and were actually achieving their original objective and reaching Heaven with their tower. God, however, had other plans for the people of Babel and made each of them speak a different language, thus rendering communication impossible. Rather than try to solve this problem, the Babelonians decided to go their separate ways, with the result that the world is now made up of people who speak different languages and the tower, sadly, was never completed.

There are many advantages to having a common language, including:

- Saving on training. If an organisation uses three or four different languages to represent different areas of business, clearly it is necessary to train staff in all languages. By using a single, unified language, it is possible to save by training all staff once in the same language. This also means that the company knowledge of that language will increase and improve as every member of staff will be familiar with the language.

- Cross-boundary communication. If people from different areas of business or departments in a company all speak the same language, clearly the communications between such people will improve and leave less room for ambiguity.

A common language also equates to having a common medium to communicate over and thus a common ground for everyone to use.

1.2.5   A good approach

In order to address the three evils of systems engineering, it is important that any project is approached in a concise, well-understood and well-defined manner. If the approach that is adopted does not take these three evils into account, the three most common causes of project failure and disaster are left unchecked and the battle to deliver a project on time will be lost almost before it is begun. Such an approach, therefore, has the following requirements:

- Consistent. The approach must be correct with regards to itself and must make sense to those adopting it. Many approaches that are defined in books, standards and other forms of reference are actually not correct, in that they contain inconsistencies or do not have a common vocabulary. If this is the case, there is a strong possibility that people will not understand the approach properly and this will lead to communication problems and complexity.

- Concise. Any approach chosen must contain enough information to convey all of its meaning, yet not so much that it becomes verbose. One of the major problems facing anyone who is trying to introduce a new approach, or to standardise an existing approach, is to get it into a form that is manageable and that will deter potential users by its sheer volume and complexity.

- Repeatable. In order to perform any type of worthwhile assessment on a project, it is important that the approach is repeatable. This means that results from more than one project may be compared and value drawn from these comparisons. If the approach is required to be improvable in some way, then making sure that it is repeatable is absolutely critical.

- Measurable. Having a concise, consistent and repeatable approach is still not enough unless the outputs of the approach (the things that are delivered on the project) are measurable in some way. Again, if the approach is to be improvable, then being able to measure it is essential.

- Improvable. This is important in order to ensure that the approach is as good and reliable as possible with sustained implementation over a long period of time. This is a point that has already been mentioned twice and is one that feeds into two of the other requirements – those of repeatability and measurability.

- Tailorable. No single approach will be applicable to every type of project and therefore it is essential that any defined approach must be 'tailored' to meet the needs of an individual project.

From this list of requirements, it should be clear that there is far more work involved in creating an approach than meets the eye. This will be a recurring theme throughout the rest of this book and is the focus of Chapter 6.

1.2.5.1 Summary

In summary, therefore, the following points were raised in this section:

- Complexity will always exist in systems as either essential complexity of accidental complexity. Whichever form exists, it is crucial to identify all areas of potential complexity and, once identified, to address them in some way.

- Communication or, more to the point, effective communication, is key to the success of any project. One crucial part of this is to have a common language that can be adopted by all people involved in the project.

- Above all, a well-defined approach is required – one that can be understood by all people involved in the project.

These points raise issues that relate directly to 'quality', which will be discussed in detail in the next section.

1.3 Quality

1.3.1 Introduction

There is a need for quality throughout today's society. This quality is required for everything that people may buy, be it a product or a service. As engineers, it is important to have a good idea of what quality is and what it means to us as professionals. Apart from a moral and ethical obligation to produce quality systems, there is also a legal obligation, in many cases, to be able to demonstrate the quality of a product. An example of this is if, as a practising engineer, you produce a piece of software that results in the death or injury of another person, it is the engineer (you) who are personally responsible, rather than the company employing the engineer. It does not matter if there are disclaimers on the software (which so many manufacturers seem keen to put on their products) because the moment a person is hurt, all disclaimers mean nothing! The only way that it is possible to prove innocence under such circumstances is to demonstrate, in a court of law, that current best practice had been followed to produce the software or, put another way, you must demonstrate the quality of the approach to produce the code [10, 11].

It is crucial, therefore, that engineers provide products and services in which people can have a high degree of trust and confidence. There are several ways that, as a customer, it is possible to judge or assess the quality of the service or product. Consider, for example, purchasing any product for use by your family. Imagine that you were about to buy a car seat for a baby or small child for use in a family car. Clearly, the product that is bought must inspire a high degree of confidence from the customer's point of view as, in the worst-case scenario, there is a child's life at risk if something goes wrong.

It is crucial, particularly in the case of children, that the product is safe and has met stringent quality guidelines, but the problem is, how does one know that the product has met these guidelines? In real life, the first thing that should be looked for would be an indication of the quality of the product. In the United Kingdom, for example, one would look for the Kite mark, which is an indication that the product has met a particular British Standard. Indeed, one would hope that if a product such as a child's car seat did not display a Kite mark then you would not purchase that product. Another example of such a symbol that indicates quality is the European quality mark, which is depicted by the 'CE' symbol.

Any product that displays such a quality symbol gives the customer confidence in the product and assurance that the product has passed stringent safety and quality tests. The same is true for services, not just products. When employing a person or awarding a contract it is usual to look for certain indications of quality of work, such as: examples of previous work, qualifications, experience and professional status (for example, membership of a professional organisation, chartered status, etc).

Certain manufacturers or service providers have such a good reputation that simply seeing their logo or trademark will inspire confidence in a product or service. Think of car manufacturers and then think of the following criteria: safety, comfort, build quality, performance and budget. When looking at each of these criteria, it is likely that particular car manufacturers will spring immediately to mind.

However, can the same be said of the way that we approach systems engineering? What should people look for when purchasing or specifying systems?

This section addresses this point in particular and discusses how to demonstrate quality of both products and services and relate them to systems engineering. It also derives a number of definitions for terms such as quality that are crucial to our understanding of systems engineering.

1.3.2 The need for quality

The term 'quality' is frequently used and quoted, yet is often misunderstood. Before we can understand quality, it is important that we understand the need for quality. There are many reasons why people require quality for their organisation:

- Customer confidence. If the customer has confidence in a product or brand of product, they are likely to select these products over others. Certain organisations are always associated with a particular level of quality of product and, given a choice, customers will almost always select the quality brand. Consider the example of buying a brand new car. Which manufacturer will you choose? Clearly, money is a major driving factor behind the purchase of a car, but what about the actual name of the manufacturer? How much impact will that have on the decision to buy or not to buy the particular brand of vehicle?

- More maintainable. Systems that are more maintainable will save money over those that are not very maintainable. Many examples of this can be found in current literature. For example, experience has shown that the most expensive phase of the life cycle is the maintenance phase. Systems should be designed with maintainability in mind.

- Common sense. At the end of the day, it is simply common sense to provide quality systems.

Unfortunately, it is a fact that many organisations will not act unless it either saves money or makes money. It is therefore very important to be able to relate quality back to financial terms as, above all else, this is (unfortunately) the only way that some individuals or organisations will react. It should be made quite clear how each of the points raised here may relate directly back to finance.

1.3.3 Level of responsibility

The level of responsibility taken on by complex systems is constantly increasing. As these systems permeate our lives, often in ways that go unnoticed, they are being given more and more responsibility, especially where safety-critical systems are concerned. Areas in which complex systems play a role that includes safety-critical elements include:

- Closed loop control systems. Today, some computer-based systems are completely in control of a number of areas of process control, such as manufacturing processes and plant monitoring processes. It is essential, therefore, that any such control systems have been designed and built with safety and reliability in mind.

- Autopilots for aeroplanes and ships. Autopilots are increasingly being used in modern transport. Many air passengers clap after the successful landing of an aeroplane, but in most cases they are clapping a piece of software, as humans are mainly required for emergency or atypical situations. Needless to say, the software has no appreciation of such applause although it does amuse pilots!

- Braking systems in cars. As technology is used increasingly in cars, so we see an increase in the level of responsibility that these complex systems have. The modern car may have computer-controlled brakes, suspension and steering, in addition to non-safety-critical applications such as electric windows, electric wing mirrors and windscreen wipers. Many cars today have cruise control, systems to aid steering and even systems to aid more complex tasks such as parking.

This increased level of responsibility thus increases the need for quality in the systems and the way in which they are produced.

1.3.4   Requirements for systems professionals

Over the last few years, the requirements for systems professionals have changed. It is no longer good enough to be a programming guru or an extraordinary designer. Instead, it is becoming increasingly important to show an awareness of wider issues, even if you are not an expert in them. These include:

- Quality policies, standards and procedures, both in-house and those in the public domain. Any member of an organisation who is not aware of quality issues, even at a high level, runs the risk of causing problems on a project. Even if no problems arise directly from this, it is highly possible (and, indeed, probable) that an organisation will fail a quality audit if members of their staff are not aware of the company's quality policies.

- Safety policies. These may be policies that relate to the system itself, such as safety-critical systems, or they may be basic safety policies for staff, such as fire regulations. Clearly, failure to meet either of these could result in injury or loss of life.

- Roles and responsibilities. It is essential that every task or action carried out by an organisation has a role defining who is responsible for that action. This is particularly important when things go wrong, or if things 'don't get done', as it provides a good starting point for getting to the heart of the problem. For example, who takes the blame when a system developed by you causes an accident? This has already been mentioned and, if appropriate roles and responsibilities are not defined, could mean blame for something arriving on your shoulders, which, by rights, should not be there.

These other areas of interest of which engineers must have a knowledge are known as processes. Therefore, it is important for engineers to have a high-level knowledge of processes that may be regarded as beyond an engineer's sphere of interest.

1.3.5   Processes and quality

The points raised in the previous section are all well and good, but how does this relate to quality? In order to answer this, it is time to come up with some definitions for quality. This is very difficult as there are many, many different definitions for quality. The two definitions chosen here are as defined by ISO (International Standards Organisation) and are:

- 'fitness for purpose'.

- 'conformance to requirements'.

'Fitness for purpose' means that the system must do what it is supposed to do. This is actually not the same as saying that the system should work. Proving that a system works is 'verification', while proving that something does what it is supposed to do is 'validation'. Therefore, it is possible to have a fully working system that has been verified, but if it does not meet its original requirements it cannot be validated.

'Conformance to requirements' means that the system must be able to be demonstrated to meet its original requirements. The original requirements state what the system is supposed to do and thus conformance to requirements is really another way of saying 'validation'.

In order to demonstrate something, it is vital to have a source of reference against which the demonstration can be compared. In the case of demonstrating quality, the source of reference that is compared against is usually a standard, a procedure, or both.

When demonstrating quality, it is the approach taken that is compared to an established norm in order to demonstrate compliance. Formally, this approach is known as its 'process'. Processes have been encountered twice already in this chapter in the following way:

- Requirements for a good approach were drawn up, which means that they are requirements for a process.

- A sound knowledge of processes was required for systems professionals, which means that these processes describe an approach to engineering, management, support services, customer supplier relationships and organisational activities.

Therefore, it is the process that is assessed or audited when demonstrating compliance with a standard, which will demonstrate the quality of an approach.

1.3.6   Standards

An important aspect, therefore, of demonstrating quality is to have a standard according to which compliance can be demonstrated. The definition of the 'standard' as used here is 'a form of reference that has been agreed upon'.

Unfortunately, in the world of systems engineering, there are literally hundreds of standards, which is simply too many! Interpreting this according to our definition, there are a great many references that have been agreed upon!

In order to make sense of this incredible number of standards, they may be grouped into three broad categories:

- International standards, which are recognised in many countries, such as: ISO (International Standards Organisation), IEEE (Institute of Electrical and Electronic Engineers), IEC (International Electrotechnical Commission), IEE (Institution of Electrical Engineers), BSI (British Standards Institution), EN (European Normative – except in French) and ANSI (American National Standards Institute). Although some of these appear to be country-specific, such as BSI and ANSI, they are in fact recognised in many countries. Most companies will comply, or try to comply, with at least one international standard. These are usually the standards that give a basic quality indicator recognised in many countries.

- Industry standards, which are defined when a group of industrial companies get together and decide to do things in the same way. Industry standards include: UML (Unified Modelling Language), CORBA (Common Object Request Broker Architecture) and ODBC (Open Data Base Connectivity). At least some of these will be familiar to most readers (if not now, then by the end of this book). The rationale behind this is that international standards are typically very slow to produce and many never make it past the draft stage before they are abandoned.

- In-house standards, which are defined on a company-by-company basis and are found in all large organisations. Examples of these are available in the public domain and include the European Space Agency (ESA), which publishes its software standards in a book, and DERA (Defence Evaluation Research Agency (whose name is apt to change), which is UK-based) standards, which are available on the Internet.

Perhaps it is due to the large number of standards that it is very difficult to understand many of them in any practical fashion. The next section looks at a number of the problems faced by anyone wishing to use any of the standards.

1.3.7   Problems with standards

Many problems exist with standards, which is particularly annoying as much of this chapter has so far been devoted to stating how important they are for effective systems engineering. There are two main areas into which these problems fall, which are problems within individual standards and problems between different standards.

- Some standards are too long. For example, both ISO 15504 Software process improvement and capability determination (SPICE) and EIA 731 Standard for systems engineering capability are several hundred pages long. The problem here is that many potential users will be deterred even before they begin to try to understand and use the standards. However, length is not always superfluous, nor is it an indication of complexity as in the case of the two standards mentioned here – the information in the standards is very clear and, quite unbelievably, very easy to understand. The length of the documents stem from the fact that they are simply very thorough and both attempt to provide a framework for assessing every process within an organisation with regards to process improvement.

- Some standards are too brief and, hence, are too high level to be of any practical use. As an example of this, ISO 9001 (arguably the most widely recognised standard in the world) is only 24 pages long. This is a standard that covers almost all applications of systems specification and design, yet it is incredibly brief. As a consequence, the standard itself is written at a very high level and is thus open to a great deal of ambiguity when being interpreted by different people. The balance between a standard being too long or too brief is a difficult concept to achieve.

- Some standards are very difficult to understand. This may be for any number of reasons, such as: they are too long or too brief, as mentioned previously, or they have been written by a committee. Of course, they may simply be badly written!

The points raised so far are relevant when a single standard is being used; however, in many cases, it may be desirable to meet more than one standard. Consider the case where a company may have an organisation-wide policy to meet an established international standard such as ISO 9001 Model for quality assurance in design development, production, installation and servicing, but may also have to meet an industry-specific standard such as EN 50128 Railway applications, software for railway control and protection systems, which applies to the rail industry. There may also be government standards that have to be met, such as, in the case of the rail industry in the United Kingdom, the guidelines laid out by HMRI (Her Majesty's Railway Inspectorate).

There are problems associated with trying to meet more than one standard that are in addition to all the points raised concerning single standards:

- Nomenclature. The terms used by different standards differ enormously, with some basic concepts having completely different meanings. This comes back to communications problems as each standard is written in a different language.

- Some standards assume that particular processes may already exist or have already been defined or complied with. A number of standards, for example, assume an ISO 9001 standard-compliant management system. This adds additional constraints to what may at first impression seem like a single standard and will include all the problems mentioned previously in this section.

Meeting a single standard is, in many cases, difficult to demonstrate. Meeting more than one standard is often orders of magnitude more difficult than meeting a single one.

1.3.8   Consequences of problems with standards

All of the problems raised in the previous section will lead to adverse affects on the project. These will include:

- Complexity. If not identified and addressed, this may lead to communication problems if complex information is conveyed to other people and, hence, to a lack of understanding.

- Communication problems. These may lead to complexity if ideas are not communicated properly by a team and, hence, to a lack of understanding.

- Lack of understanding. This may lead to communication problems if the information is not fully understood and, hence, complexity if the information modelled is not correct, concise or consistent.

These three points should be quite familiar by now and should only serve to reinforce how important it is that these three evils can be addressed.

1.3.9   Compliance with standards

Compliance with a standard is generally demonstrated by carrying out either an assessment or an audit. These two are similar yet subtly different and, in very simple terms, can be differentiated as follows:

- An audit. During an audit, an organisation has its practices compared to that of a standard as defined and documented in the organisation's operating procedures. This often takes the form of taking these procedures as a starting point and then actively looking for discrepancies or non-compliances with the standard. Formal audits, which result in certification according to a standard, are carried out by independent, third-party organisations. The result of an audit is a simple 'yes/no' or 'pass/fail' with minimal information concerning the non-compliances, usually in the form of a simple reference to the relevant section in the standard.

- An assessment. During an assessment, an organisation has its processes compared to those of an established process model, usually a standard. An assessment starts with a blank sheet of paper and whatever processes are executed (whether defined and documented, or even assumed) are assessed. Assessments may be carried out independently or in-house, providing that the assessor is qualified in process assessment. The result of an assessment is a 'capability profile', wherein each profile that has been assessed is rated according to a pre-defined scale, which provides an indication of how 'mature' each process is.

In order to demonstrate compliance with a standard, there are a few basic requirements that must be met:

- A defined process must exist. Without a defined process, it is impossible to meet any type of standard with any degree of satisfaction as this is usually the thing being assessed in an audit for example. By 'defined', it means that the process must be documented in some way, such as a formal document or electronic copy that is available to staff (such as an Intranet or Internet-based definition).

- Use of the process must be demonstrated. It is not good enough simply to have a defined process if nobody follows it. It is important to be able to show examples of the processes in action or, to put it another way, projects. The things that are evaluated in order to demonstrate the use of the process are the process deliverables, or artefacts. These deliverables may be documents, code, hardware, reports, etc.

- The process must be able to be improved. A defined and implemented process is just a first step. It must be shown that the process works effectively and that mechanisms are in place to continuously improve the process as time goes on.

One approach to demonstrating standard compliance with a standard is to represent both the process and the standard using a common language, which will overcome many of the communication problems. Once both the process and the standard are represented in a common language, it is possible to compare them directly. This approach is also applicable to the case where more than one standard is required for compliance.

1.3.10 Processes

In very simple terms, a process describes an approach to doing something. It is something that transforms inputs into outputs and has responsibility associated with it.

Everybody actually follows a process whenever they do anything, even if they fail to realise it and it does not even have to be written down. When people do things in a 'certain way', they are actually following a process, albeit an informal one. Thus, if we want to demonstrate the quality of the way in which we work, it is important to be able to identify the different areas of activity that relate to work. It is important, therefore, that key processes in an organisation can be identified and defined in a way that everyone can understand. This relates directly back to the point made previously concerning people having a greater understanding of the way in which a company works. In real terms, this equates to staff having a high-level knowledge of the types of process (that describe an approach to work) that exist within their organisation. Typical processes within an organisation include:

- Customer supplier processes, which include protocols and a defined way of dealing with customers and for customers to deal with suppliers.

- Support services, which represent services such as change control, configuration management, etc.

- Management, which includes processes such as project management and project planning, which apply on a project-by-project basis rather than across the whole organisation.

- Organisational processes, which will be applied across the whole enterprise, such as writing standards, health and safety, etc.

- Engineering, which covers the development of a product from requirements to retirement.

It is important that these processes are identified and categorised in order to make understanding them as simple as possible. The categories introduced here are based on the ISO definitions of processes, but are by no means carved in stone. In fact, Chapter 6 shows several other categorisations of processes according to different standards.

1.3.11 Summary

In summary, therefore, the following points have been raised in this section:

- There is a need for quality in everything that we, as systems professionals, do and produce.

- Quality is most often demonstrated by compliance with a standard. However, many problems exist with understanding such standards, either individually or across more than one standard.

- In order to demonstrate compliance with a standard, a well-defined process is essential.

Processes are discussed in more detail in Chapter 6, where an approach to solving the problems raised here is suggested. Several real-life examples of modelling standards are given, together with an example process that is defined according to these standards.

1.4 Conclusions

This chapter has raised several of the main issues that affect systems engineering in today's world. In summary, the key points are:

- Project failures and disasters are commonplace in today's technology-driven society. The result of such failures and disasters can range from loss of money, to loss of life and damage to the environment.

- Some of the common themes behind many such failures and disasters are the three evils of systems engineering: complexity, communications and a lack of understanding.

- In order to control and minimise these problems, it is essential that the way we work is of the highest quality and that this quality can be demonstrated to other people. As quality was defined as 'fitness for purpose' and 'conformance to requirements', it is these two definitions that must be demonstrated. In practical terms, this translates to having a well-defined process.

    Finally, two definitions of systems engineering were suggested:

- "Systems engineering is concerned with defining and implementing an approach to solving problems, while managing complexity and communicating effectively over the entire lifetime of a project".

- "Systems engineering is the implementation of common sense".

    The information in this chapter will form the basis for the remainder of the book.

1.5 Further discussion

1. Think of the following terms and see if you can associate a car manufacturer with the term: 'expensive', 'safety', 'reliability', 'expensive to maintain', etc.

2. Is ISO correct in its definition of the five main categories for processes within an organisation? Do these adequately represent the way that you or your organisation work?

3. Look out for quality symbols on products and services that you encounter. Find out which standard they comply with and check this standard to see what it actually means to you as a customer.

1.6 References

1   STEVENS, R., BROOK, P., JACKSON, K., and ARNOLD, S.: 'Systems engineering, coping with complexity' (Prentice Hall Europe, 1998)

2   O'CONNOR, J., and McDERMOTT, I.: 'The art of systems thinking, essential skills for creativity and problem solving' (Thorsons, London, 1997)

3   SCHACH, S. R.: 'Software engineering with Java' (McGraw-Hill International Editions, 1997)

4   BIGNELL, V., and FORTUNE, J.: 'Understanding systems failures' (Open University Press, 1984)

5   COLLINS, T., and BICKNELL, D.: 'Crash, ten easy ways to avoid a computer disaster' (Simon and Schuster, 1997)

6   FLOWERS, S.: 'Software failure: management failure, amazing stories and cautionary tales' (John Wiley and Sons Ltd, 1997)

7   LEVESON, N. G.: 'Safeware, system safety and computers' (Addison-Wesley Publishing Inc., 1995)

8   LEWIN, R.: 'The major new theory that unifies all sciences, complexity, life on the edge of chaos' (Phoenix Paperbacks, 1995)

9   BROOKS, F. P.: 'The mythical man-month' (Addison-Wesley Publishing Inc., 1995)

10 AYRES, R.: 'The essence of professional issues in computing' (Prentice Hall, 1999)

11 BAINBRIDGE, D.: 'Computer law' (Pitman Publishing, 1996)

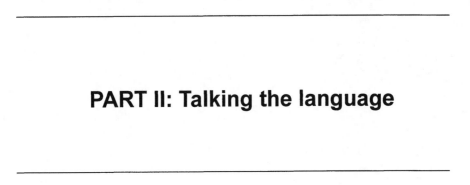

PART II: Talking the language

Chapter 2

Modelling

"better to understand a little, than to misunderstand a lot"
– Homer Simpson

2.1 Introduction to modelling

This section introduces the concept of modelling and includes a discussion on why modelling is so important and why we need to model. It also contains information that forms the basis for the remainder of this part of the book and introduces a number of basic concepts that will be referred back to constantly throughout the rest of the book.

2.1.1 The importance of modelling

It is vital to understand the reasons why we must model things. In order to justify the need for models, it is probably easiest to look at a number of simple examples, based on real world systems, to which people can relate.

Note that here we are justifying modelling in general terms and not simply with regard to software systems, which demonstrates the need for flexibility when modelling – indeed, the three examples used here are non-software systems. Many examples are used in many different books, but the examples used here are based on those defined by the master of modelling, Grady Booch [1]. There seems little point in contriving a new modelling example when there is a perfect set of examples that has already been defined and is widely accepted as the norm. Therefore, the examples chosen here are based on Booch and his doghouse, house and office block.

2.1.1.1 The kennel (doghouse)

For the first example of modelling, consider the example of building a kennel. A kennel is a small house where pets, usually dogs, can spend some time outside without being exposed to the elements. The basic requirement for a kennel is to keep the dog happy. This will include building a structure that is waterproof and large enough to fit the dog inside. In order to fit inside, there must be an entrance in the kennel that is, preferably, larger than the dog itself. The inside should also be large enough for the dog to be able to turn around in order to leave. Dogs are particularly bad at walking backwards, which makes this last point crucial. Finally, the dog should be comfortable enough to sleep in the kennel and thus some bedding or cushions may be in order.

If you were about to build this kennel, then consider for a moment the basic skills and resources that you would require. You would be wise to have:

- Basic materials such as timber, nails, etc. The quality is not really that important as it is only for a dog. This might involve looking for any old pieces of wood around the house or even making a trip to a hardware store.

- The money needed to pay for the kennel would be your own, but is unlikely to be a large outlay. In terms of your personal income, it would be a fraction of a week's salary – perhaps the cost of a social evening out.

- Some basic tools, such as a hammer, a saw, a tape measure etc. Again, the quality of the tools need not be wonderful, providing they get the job done.

- Some basic building skills. You need not be a skilled craftsman, but basic hand-to-eye co-ordination would be an advantage.

At the end of the day (or weekend), you will probably end up with a kennel that is functional and in which the dog would be happy to shelter from the rain.

If the kennel was somewhat less than functional and the dog was not very happy with its new accommodation, you could always start again (after destroying the first attempt) and try a bit harder, learning from past mistakes. Alternatively, you could get rid of the dog and buy a less demanding pet such as a tortoise, as there is no need to build a kennel for an animal that carries its own accommodation on its back. After all, the dog is in no position to argue or complain.

This is Booch's first example of modelling: the kennel or doghouse.

2.1.1.2   The house

Consider now, maybe based on the resounding success of your kennel, that you were planning to build a house for your family. This time the requirements would be somewhat different. There would need to be adequate space for the whole family in which to sit and relax. In addition, there would have to be adequate sleeping arrangements in the number of bedrooms that are chosen. There would need to be a kitchen, maybe a dining room and one or more bathrooms and toilets. As there will be more than one room, some thought should be given to the layout of the rooms in terms of where they are in the house and where they are in relation to one another.

If you were to build a house for your family, you would (hopefully) approach the whole exercise differently from that of the kennel:

- You would have to start with some basic materials and tools, but the quality of these resources would no doubt be of a higher concern than those used for the kennel. It would not be good enough to simply drive down to a local hardware store and pick up some materials as the quantity would need to be far greater and it would not be possible to guess, with any degree of accuracy, the amount of materials required.

- Your family would also be far more demanding and vocal than the dog. Rather than simply guessing your family's requirements, it would be more appropriate to ask them their opinions and perhaps get a professional architect in to listen to and discuss their needs.

- Unless you have built many houses before, it would be a good idea to draw up some plans. If you were hiring skilled craftsmen to do the job, you would certainly have to draw up plans in order to communicate your requirements to the builders. These plans may require some input from an architect in order that they achieve a standard that may be used effectively by the people who will be building the house.

- The house would also have to meet building regulations and require planning permission. This may involve contacting the local council or government representative and possibly applying for permission to build. This in turn would almost certainly involve submitting plans for approval before any work could be started.

- The money for the house would probably be yours, and thus you would have to monitor the work and ensure that people stick to the plans in order to get the job done in time, within budget and to meet your family's original requirements. The scale of the financial outlay is likely to be in the order of several years' salary and would probably be borrowed from a bank or building society and would thus have to be paid back, regardless of the outcome of the project.

If the house turns out not to suit the requirements, the consequences would be more serious than in the case of the kennel. The house cannot be knocked down and started again as the kennel could, because considerably more time and money would have gone into the house building. Similarly, you cannot simply get a less demanding family (in most cases) and living with the consequences of failure is not worth thinking about!

This is Booch's second example: the house.

2.1.1.3 The office block

Taking the two building projects that have been discussed so far even further, imagine that your ambition knows no bounds and that you decide to build an entire office block.

Consider once more the resources that would be required for this, the third and final building project.

- It would be infinitely stupid to attempt to build an office block by yourself.

- The materials required for building an office block would be in significantly larger quantities that the house (and most definitely the kennel). The materials would be bought direct from source and may even need to be brought in from specialist suppliers, perhaps even in different counties or countries.

- You will probably be using other people's money and thus the requirements for the building will probably be their requirements. In addition, their requirements will no doubt change once you have started building the tower block.

- More permissions are required to build an office block than a house and many more regulations must be considered. Consider, for example, environmental conditions that the office building may have to meet – the building must not block anyone's light, it will be required to blend in with its surroundings, or it may be deemed too ugly for a particular area.

- You will have to carry out extensive planning and be part of a larger group who are responsible for the building. Many teams will be involved from different areas of work (builders, plumbers, electricians, architects, etc), all of whom must intercommunicate.

If you get the right teams involved and enjoy a degree of luck, you will produce the desired building.

If the project does not meet the investor's requirements, you would face severe repercussions, including the potential of no further work and the loss of reputation.

This is Booch's third example: the office block.

2.1.1.4 *The point*

These three examples from Booch may seem a little strange and somewhat trivial at first glance; however, there is a very serious and fundamental point behind all of this.

Nobody in their right mind would attempt to build an office block with basic DIY skills. In addition, there is the question of resources, and not only in terms of the materials needed. In order to build an office block, you would need the knowledge to access the necessary human resources (including people such as architects, builders, crane operators, etc), plenty of time and plenty of money.

The strange thing is that many people will approach building a complex system with the skills and resources of a kennel-builder, without actually knowing if it is a kennel, house or office block. When contemplating any complex system, you should assume that it will be, or has the potential to turn into, an office block building. Do not approach any project with a 'kennel' mentality. If you approach a project as if it were an office block and it turns out to be a kennel, you will end up with a very well-made kennel that is the envy of all canines. If, however, you approach a project as if it were a kennel and it turns out to be an office block, the result will be pure disaster!

One of the reasons why it is so easy to misjudge the size and complexity of a project is that, in many cases, many elements of the system will not be tangible or comprehensible. Consider integrated circuits: who can say, simply by looking at an IC, how many transistors are inside it? Is it a processor or a simple logic gate? Consider software: simply by looking at a CD-ROM it is impossible to judge the size or complexity of the information contained on it. In terms of size, this could range anywhere from a single kilobyte to several Gigabytes (if the information is compressed or 'zipped'). Even if you have an idea of the type of application that is on the CD-ROM, it is still difficult to judge the size of the software. Take for example, software that may help people write letters: this may be a simple line editor program that may be a few kilobytes of information or it could be a full-blown office application that may not fit on to a single CD-ROM.

The important term that is used here is 'complexity' rather than size, as size is not necessarily a reflection of the complexity of a system.

2.1.2   Why projects fail

Projects fail for many different reasons. However, there are several underlying reasons, or themes, that tend to emerge when looking at project failures and disasters in general. There are three main themes, which have already been discussed in some detail in Chapter 1:

- Complexity. Complexity is a huge problem. Some complexity is due to poor engineering or the second and third reasons on this list, which is known as 'accidental complexity'. Accidental complexity can be avoided or, at the very least, should be minimised. Unfortunately, there is a second type of complexity, 'essential' complexity, which is unavoidable as it is in the essence of the system.

- Lack of understanding. Software systems are applicable to a great many types of system. Unfortunately, software engineers do not understand all different types of system, and people, generally, do not understand software.

- Communication. Communication, or rather a lack of it, is a problem that occurs in all aspects of life. People may not be able to communicate effectively because they speak different languages, they have different areas of expertise, they have different definitions for the same term, or they cannot describe their thoughts effectively.

However, many projects do succeed. One reason is that they avoid, or minimise, the aforementioned problems due to effective modelling.

2.1.3   Modelling

Are there any other reasons why we model?

The three points that were raised in the previous section have already been discussed and form one of the basic requirements for modelling. After all, if these are the three biggest problems with systems and modelling is one of the reasons why projects succeed, it is not unreasonable to relate the two together.

Other requirements for modelling emerge from the three themes mentioned previously. There is a need, for example, to visualise systems as they will appear as a final product. Reconsider, for a moment, the three examples already introduced in this chapter: the kennel, the house and the office block. Almost anybody should be able to distinguish between the three types of building, even if they are shown different representations of the same type of building. This is because when we hear a term, such as house or office block, we can immediately visualise what the system will look like. However, when we hear other terms, such as automated control system or ecosystem, it is unlikely that any two people would come up with a similar image of what the final system would look like as they are not physical systems. It is therefore important that we can visualise the final system at the outset to give some idea of the usability and functionality of the system. This may include building a small physical model of a building, drawing a diagram of what something should look like, etc. This helps potential users identify with what they will be receiving. It would, after all, be very foolish to pay for an office block when you had not seen any plans or models of what the final building would look like!

These plans and models may also be used for another purpose – that of analysing a final product before it is built. This may be something very simple, such as getting the colour or size right, or it may be something far more complex, such as stress analysis, wind-tunnel testing, performance loading, and so forth.

Models may also be used as a template for creating or constructing the final system. This is simple to understand within the context of building a house as the house is based directly on the plans that are drawn up by the architect. The same is true for circuit designs, which are simply models, albeit in an abstract form, of the circuit that is to be built.

In order to understand one final use for models, it is necessary to take a step back and consider what happens in a project when something does go wrong. Typically, what will happen is that someone will ask why a particular design had been adopted. By modelling throughout a project it is possible to model, not only a single solution, but also a set of candidate solutions that may be assessed and a final solution chosen that best meets the original requirements of the system. These models of the complete set of candidate solutions may then be used as a basis for documenting decisions and if the models are retained so is the decision-making process. This is useful for justifying why a particular approach or solution was chosen over any other choices. This is particularly useful when people are trying to 'pass the buck'.

After all these discussions concerning the rationale for modelling, it is now time to define exactly what a model is.

2.1.4   Defining a 'model'

It is now possible, bearing in mind the previous discussion, to come up with a definition for a model, which will be taken once more from Booch.

> *A model is defined as a simplification of reality that is created in order to better understand the system under development, as we cannot comprehend complex systems.*

We will return to this definition throughout this chapter and the rest of the book and thus it should be kept at the forefront of your mind while reading the book.

In summary, therefore, we use modelling in four explicit ways:

- To visualise a system.

- To specify a system. By specify, we mean to state specific needs or requirements concerning the system.

- As a template for creation.

- In order to document decisions made throughout the project.

These four aims of modelling are used to help to reduce complexity, improve communications and enhance our understanding of a system.

2.1.5   Principles of modelling

Booch identifies four principles of modelling that are deemed crucial for successful and consistent modelling: the choice of model, the level of abstraction, connection to reality and independent views of the same system.

2.1.5.1   The choice of model

The choice of model will have a profound influence on how a problem is approached. Approaching a problem the right way can make a job much simpler and will be quicker than adapting the wrong approach.

Consider for example other subjects that are taught in a school or college. Many subjects are taught by advocating a particular approach to analysis or problem-solving but an inherent part of the approach will be to choose the right way to solve the problem. As an example of this, consider a simple mathematical operation, such as finding the slope of a line. This may be done in many ways, depending on the type of line. If it is a straight line, a simple 'height-over-length' equation will solve the problem. However, if, on the other hand, the line is a curve, this simple formula will not work, as the slope of the line will change depending on the point in the line at which the slope is measured. In order to solve this problem, calculus and in particular differentiation should be applied. Applying this retrospectively, differentiation can be applied to a straight line with the same result as the simple formula, but not vice versa.

Everybody must have sat in an examination and started out answering a question, only to find that, halfway through, the wrong approach has been taken and the problem must be restarted. This could have a big influence on your life, particularly if you do not finish the exam or if you get the answers wrong. Also, remember, as some people are so fond of saying: you get more marks for your approach than for the actual answer! It is the same with all engineering, a well-defined approach is vital, which involves choosing the appropriate model.

This is a topic that will be raised many times once the UML has been introduced.

2.1.5.2   Abstraction of the model

For the purposes of this book, 'abstraction' refers to the level of detail of a model. Imagine the office block problem. You may need to create a model to show the size of the whole building, you may need to create a model to show how a single storey is constructed, or you may need to create a model to show how a single light switch operates. Each of these requires a different level of abstraction for a successful model.

The point to be made here is that any model will require different levels of abstraction to be represented, otherwise the model will have little chance of being correct. Imagine the scenario in which a single model was drawn up to represent the office block. Although useful for the financial backers to see a high-level view of the building, it would be next to useless for an electrician, plumber or ventilation engineer.

2.1.5.3   Connection to reality

One problem with modelling is that, according to our definition, models simplify reality. This means that some information must be lost somewhere along the line, which can cause problems. Therefore, it is vital to know both how the model relates to real life and how far it is divorced from real life.

From a practical point of view, initial models tend to have quite a loose connection to reality and, as these models evolve, they get closer and closer to reality. The final connection to reality will be at the point when the model actually becomes reality, which is when the model is implemented or constructed based on the model.

From a conceptual point of view, it is important that whoever is looking at a model can make a connection to reality, otherwise the model becomes meaningless. Consider again the example of differentiation. School children often find differentiation difficult to understand as they cannot 'see the point' of it. Only when practical examples can be shown for using differentiation, such as working out speed and acceleration, can people truly understand the concept.

2.1.5.4 *Independent views of the same system*

It has already been stated that a good model requires views that represent different levels of abstraction, but it is also true that a good model requires views modelled from different vantage points.

Consider again the office block building where many different views will be required to completely model the system. The different people who work on the project will require different views of the same model. For example, electricians should really have wiring diagrams, while painters will require colour charts. There is no point giving the colour chart to the electrician, nor the wiring diagram to the painter, as they are simply not relevant.

One crucial point that must be made here, however, is that each of these independent views must be consistent with one another, or, to put it another way, they must integrate correctly.

2.2 The UML

2.2.1 *Introduction*

The previous sections have identified many requirements for modelling and have justified why modelling is so important. In order to model successfully, a common language is required and that common language is, quite unsurprisingly, the Unified Modelling Language (UML).

The UML is a general-purpose modelling language that is intended for software-intensive systems. This definition is far too restricting however. The UML is a language and can thus be used to communicate any form of information and should not be limited to software. The main aim of this book, therefore, is to take the UML and apply it to systems engineering, rather than simply to software engineering. There are many excellent books that relate the UML to software, so this will be kept to a minimum in this book.

2.2.2 *Types of model*

Nine diagrams exist in the UML, which may be grouped into two broad categories. These two types of model that exist in the UML are the 'static' and 'behavioural' models. It is vital that both of these models exist for any system, otherwise the system is not fully defined.

Each of the nine diagrams belongs to one of these categories. Each of the nine diagrams has strong relationships with other types of diagram and both types of model have strong relationships with one another. It is these strong relationships between the types of diagram and model that allow consistency checks to be performed on UML models, and which will give confidence in the model itself.

The static model or view of a system shows the 'things' or entities in a system and the relationships between them. In a way, the static model shows 'what' the system looks like and 'what' it does, but not 'how'. A static model may be thought of as a snapshot in time of any system.

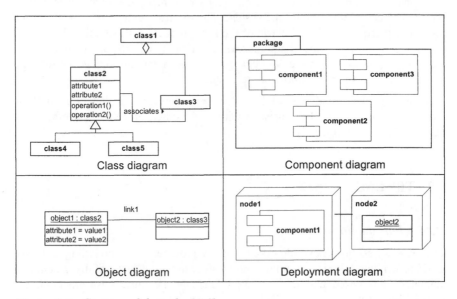

Figure 2.1   *Static models in the UML*

Figure 2.1 shows the UML diagrams that relate to static modelling. Static models may be realised by four UML diagrams: class diagrams, object diagrams, deployment diagrams and component diagrams. Static modelling is the topic of Chapter 3, while each of the types of static diagram are discussed in more detail in Chapter 5.

The static model shows the 'what' of the system, while the behavioural model shows the 'how'. Behavioural models demonstrate how a system behaves over time by showing the order in which things happen, the conditions under which they happen and the interactions between things.

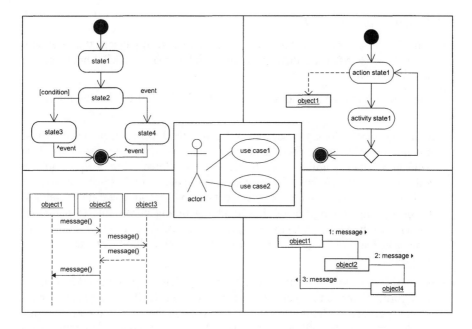

Figure 2.2 *Behavioural models in the UML*

Behavioural models may be realised by five UML diagrams: statecharts, use cases, interaction diagrams (sequence and collaboration diagrams) and activity diagrams. Behavioural modelling is the subject of Chapter 4, while each of the behavioural diagrams is discussed in more detail in Chapter 5.

2.2.3   Views vs. models

Many books refer to the different types of 'view' of a system that must exist throughout the life cycle of the project. This term is often used interchangeably with the term 'types of model', but for the purposes of this book it is important to distinguish between them. A view refers to a collection of UML diagrams that are created for a specific purpose and that are often related to different phases of the project life cycle. The two types of model that are used in this book are still applicable to the types of view that are defined in other books. The reason why these views are not discussed in much detail here is that they often constrain development to a particular process or design approach, while this book introduces the UML purely as a language and nothing more.

The UML should not change the way in which you approach your work – instead, it should complement it and make it more efficient.

2.2.4   Extending the UML

Despite what others may say, the UML is not some sort of panacea that will cure all your problems. After all, there is no silver bullet and the UML should not be considered as such.

There are several applications for which the UML will not be suitable, such as:

- Real-time applications. One criticism that is often levelled at the UML is that it is not suitable for the specification and analysis of real-time systems. This criticism is slightly unfair as the UML was never intended to be applicable to real-time systems. However, having said this, much work has been carried out on how the UML may be extended to take into account real-time systems. These techniques range from simply adding timing constraints to particular elements of the UML, to adding in completely new types of diagram that will cope with real-time requirements, while maintaining the overall umbrella of the UML.

- Formal, mathematical specification. Another criticism is that the UML is not a truly formal approach. Again, this is true, but the UML is intended to be a general-purpose modelling language with a scope that is as wide as possible. In addition, there is nothing to stop the UML being used in conjunction with more formal techniques, such as VDM or Z (formal specification languages), in order to formally specify crucial parts of a system.

The UML was designed to be as flexible as possible and, as such, has a number of built-in extension mechanisms that may be used to extend the functionality and scope of the language itself. These basic mechanisms are stereotypes, constraints and tagged values, which are discussed in detail in Chapter 8.

2.2.5   Background of the UML

This section gives a very brief overview of the background of the UML and discusses where it came from and why.

The UML is a general-purpose visual modelling language that is used to specify, visualise, construct and document the artefacts of any system. These terms have already been defined, but will be mentioned again here:

- To specify means to refer to or state specific needs, requirements or instructions concerned with a complex system.

- To visualise means to realise in a visual fashion or, in other words, to represent information using diagrams.

- To construct means to create a system based on specified components of that system.

- To document means to record the knowledge and experience of the software in order to show how the software was conceived, designed, developed, verified and validated.

The UML is very flexible and has a wide scope that reflects this. The UML may be used for:

- Any development environment. The UML is not restricted to any particular development environment or language. Although some people claim that the UML is more suited to object-oriented environments, it should be remembered that object orientation is, in one sense, a frame of mind, and that the UML is simply a language. Therefore you should not constrain yourself by thinking that the UML may only be used for object-oriented software systems, as it can be used for any system with any implementation in mind.

- Any process model. The UML is a language that, unlike its predecessors, does not have an inherent process. There is a defined process that is recommended by the authors of the UML, which is known as the Rational Unified Process (RUP). The RUP is an excellent, well-proven approach to software design, but there is no reason to suggest that this approach must be adopted in order to use the UML successfully. Indeed, Chapter 6examines process modelling and makes the point that the UML can be applied to any process. Remember, the UML should not change the way that you work, but it should make you work more efficiently.

- All life cycle phases. The UML should not only be considered for analysis and design, but may also be used effectively, at all life cycle phases, from requirements through to operations (or inception to transition, if you are an RUP person). The more that you use the UML, the more benefits will be derived from using the language. It makes sense that if you are using a particular language for requirements, the same language should be used for analysis and design. Apart from anything else, it aids traceability enormously, which from a systems engineering quality point of view is absolutely crucial to the success of a project. The UML should also be utilised after the implementation phase, where it can be used to plan and document the installation and transfer of a system and can even be used for user manuals and other documentation.

- Any application domain. The UML may be used for any application area, even those areas that do not include software development. Historically, many modelling techniques have emerged from a particular industry, which is fine if you happen to work in that industry, but if you move careers or if two separate industries have to talk to each other, problems can arise – the increasing widespread use of the UML overcomes this.

The UML represents a unification of past experience and methodologies in a single, cohesive language that can be applied to any sort of modelling.

2.2.5.1   History of the UML

The UML is relatively new, having only been fully defined since 1997. Modelling languages, however, have existed for many years in many different forms. However, there were simply too many modelling languages, methods and methodologies. One of the main reasons for modelling – that of communicating effectively – was being destroyed as, in a bizarre Babel-like scenario, everyone ended up speaking different languages when they were actually trying to solve this problem. There was thus a huge industrial need to consolidate all these techniques into a single, usable, flexible language. In addition, by making the UML a language rather than a methodology (with an inherent process), it became more generally acceptable and far more powerful than its predecessors.

The UML, therefore, was developed in order to simplify and consolidate the many existing methods which may be grouped into the two main categories of traditional techniques and object-oriented techniques.

2.2.5.2   Traditional modelling techniques

Many traditional development methods exist. The term 'traditional' here refers to the non-object-oriented techniques sometimes referred to as 'classic' techniques.

Some traditional techniques include:

- Structure Analysis and Structured Design (SASD) created by Edward Yourdon and Larry Constantine in 1977. This was considered the 'father' of all other techniques as it was the original visual modelling technique [2].

- Structured Analysis for Real-time Systems, by PT Ward and S Mellor. This approach was based on function-decomposition but with added diagrams to take into account timing relationships in a system [3].

Many other techniques exist which are based mainly on the data flow diagram approach. These techniques were widely accepted from the 1980s onwards and remain in widespread use today. For a discussion of these and other techniques, see [4].

2.2.5.3   Object-oriented modelling techniques

The late 1980s saw the advent of object-oriented (OO) techniques. Although object-oriented languages have been around since the late 1960s (Simula-67), it took 20 years for their associated modelling techniques to be developed.

These techniques include:

- Object-oriented systems analysis, by Shlaer/Mellor in 1988. This was an early analysis approach that made use of a type of object diagram, state transition diagrams, tables and domain models [5].

- Object-oriented analysis, by Peter Coad and Edward Yourdon in 1991. This approach used static diagrams that represented subjects and objects [6].

- The Booch method, by Grady Booch in 1991. One of the most widely adopted approaches and, of course, of the three main authors of the UML, the Booch approach used: class diagrams, object diagrams, module diagrams, process diagrams, state transition diagrams and timing diagrams [7].

- The object modelling technique (OMT) by James Rumbaugh, Michael Blaha, William Lorensen, also in 1991. Again, one of the most widely adopted techniques that made use of three main types of diagram: class diagrams, statecharts and data flow diagrams. The class diagrams from OMT are very similar indeed to those in the UML using much of the same notation [8].

- Object-oriented software engineering (OOSE) by Ivar Jacobson in 1992. Jacobson's main contribution to the world of OO has been the use of use case diagrams that dictated much of the OOSE approach. The use cases that are part of the UML are derived directly from Jacobson's work [9].

- CRC-cards by Rebecca Wirfs-Brock. CRC-cards represent a very pragmatic approach to applying OO techniques and are still widely used today. Indeed, some people still like to use them in conjunction with the UML [10].

Many other techniques also exist and, if you are interested in them, see [11, 12] for a detailed description and comparison of the different techniques.

Anyone familiar with any of the techniques mentioned here or any other OO technique will no doubt recognise elements of them in the UML. This is because the UML is a unification of all these techniques.

2.2.5.4 *Unification of different modelling techniques*
There were a number of early attempts to unify concepts among the various modelling techniques. Perhaps the most widely known (apart from the UML, that is) was Fusion, by Coleman et al. The Fusion method combined elements of Booch, OMT and CRC. However, as none of the original authors were involved, it is judged to be a new method rather than a unification effort.

The first real unification effort came from Rumbaugh and Booch from Rational software corporation in 1994, known as the Unified Method. The following year, they were joined by Ivar Jacobson. This work was then renamed the Unified Modelling Language – the UML.

In 1996, the Object Management Group (OMG) issued a request for proposals for a standard approach to object-oriented modelling. As it turned out, the UML was the sole submission to the OMG as it had been agreed upon by methodologists, industry and CASE (computer aided/assisted software engineering) tool vendors.

As stated previously, Rumbaugh, Booch and Jacobson worked with many other methodologists, including those responsible for the Fusion method, the result of which was a single submission to the OMG in the form of the UML. The

UML was accepted as a standard by the OMG in 1997 and responsibility for the UML is now assumed by the OMG [1].

As the UML is now a standard, the information is non-proprietary and thus there is little risk of being tied to a single vendor or supplier when adopting the UML. With the announcement of the acceptance of the UML by the OMG there came a flood of textbooks and CASE tools on the market, of very mixed and variable standards.

The list of backers for the UML is as impressive as it is long. The core team included Hewlett-Packard, I-Logix, IBM, ICON Computing, Intellicorps and James Martin and Company, MCI Systemhouse, Object Time, Oracle, Platinum Technology, Rational Software, Taskon, Texas Instruments and Unisys.

There are many textbooks on the market and many CASE tools. All of these resources vary enormously in quality and cost and it is important to be able to make an informed decision as to where to go next with regards to the UML.

2.2.6   A non-proprietary language

Perhaps one of the most crucial requirements for the UML was to make it non-proprietary so that it is not owned by a single organisation. Although the UML was developed by Rational software, it is the OMG who retain ownership and responsibility for the UML standard itself and, as such, it is in the public domain. This means that you can use any UML-compliant CASE tool to aid modelling, any company (with UML knowledge, obviously) for consultancy and training and choose any textbook from which to learn the UML. If there are any inconsistencies with the resources being used, or any queries with the UML, there is a single source of reference, which is the standard itself.

Another important aspect of the non-proprietary nature of the UML is that there is far less risk of being locked into a single vendor for tools and support. One massive risk to any project is that a single vendor is used to supply all tools and support (training, consultancy, etc) and this company either goes bust and hence, all support is gone, or doubles its training and consultancy fees overnight, leaving the customer in a very precarious situation.

2.3   Conclusions

In summary, therefore, there are many reasons why the UML is truly unified. The UML provides unification:

- Across historical methods and notations. The UML combines notation and concepts from many modelling techniques. The UML can represent most existing models as well as, if not better than, their existing techniques. Anyone familiar with an existing technique will no doubt recognise some elements of the UML.

- Across development life cycles. The UML may be used at all phases of the software development life cycle and thus the transition between phases is seamless. As the UML is used more often, so the advantages associated with its use will increase.

- Across application domains. Many existing techniques were focused in a single industry and thus had limited scope. The UML, however, is designed to be applicable to any aspect of complex software-intensive systems.

- Across implementation languages. The UML is language-independent and thus may be used for any form of implementation. This may relate to a particular software language or may not even apply to software at all.

- Across development processes. The UML may be used for any development process and, as will be seen in due course, may even be used to define such a process. Although there is a recommended process (the RUP) that is promoted by the designers of the UML, there is no obligation whatsoever to follow such an approach, as the UML is a language rather than a methodology.

Having established, therefore, that the UML is truly unified, it also has the following advantages over all of its predecessors:

- Widely accepted. The list of industrial supporters is very long (the list provided here being only a subset of full supporters) and is growing all the time. The UML is industry-driven and satisfies a real industrial need.

- Non-proprietary. This is crucial from a project risk point of view as the risk of being tied to a single supplier or vendor is minimised.

- Commercially supported. There are many tools and textbooks on the market to support UML efforts.

- Extendable. Although the UML is no 'silver bullet', it is designed to be as flexible as possible and to be extendable for specific applications. Chapter 8 discusses extending the UML in more detail.

All of these points contribute towards making the UML the most flexible and widely used modelling language in use today, which has superseded all previous techniques and notations.

The UML is the industry standard for software engineering and is being increasingly used for systems engineering. Knowledge of the UML is becoming more important, to the point of becoming essential for all systems engineers.

2.4 Further discussion

1. Think of some examples of the three types of model that were discussed: the kennel, the house and the office block.

2. Consider the office block and try to think of different types of model that would be required to construct the building. What different teams of people would be involved?

3. What levels of abstraction would be required for the house? Who would require these different levels of abstraction?

4. Consider the mathematical operations of differentiation and integration. What can they actually be used for in real life? How would you compare their connection to reality?

5. Consider visual modelling and formal mathematical modelling. Which would be the most appropriate for representing the architecture of a system and which for representing a numerical simulation of the same system?

2.5 References

1   BOOCH, G., RUMBAUGH, J. and JACOBSON, I.: 'The unified modelling language user guide' (Addison-Wesley, 1999)

2   YOURDON, E. and CONSTANTINE, L. L.: 'Structured design: fundamentals of a discipline of computer program and systems design' (Prentice-Hall, Englewood Cliffs, NJ, 1979)

3   WARD, P. T. and MELLOR, S.: 'Structured development for real-time systems' (Yourdon Press, New York, 1985)

4   SCHACH, S. R.: 'Classical and object-oriented software engineering' (Irwin, 1996)

5   SHLAER, S., and MELLOR, S. J.: 'Object-oriented systems analysis – modelling the world in data' (Yourdon Press, Englewood Cliffs, NJ, 1988)

6   COAD, P and YOURDON, E.: 'Object-oriented analysis' (Yourdon Press, Englewood Cliffs, NJ, 1991)

7   BOOCH, G.: 'Object-oriented analysis and design, with applications' (The Benjamin/Cummings Publishing Company Inc., 1994)

8   RUMBAUGH, J., BLAHA, M., PREMERLANI, W., EDDY, F. and LORENSON, W.: 'Object-oriented modelling and design' (Prentice-Hall, Englewood Cliffs, NJ, 1991)

9   JACOBSON, I.: 'Object-oriented software engineering, a use case-driven approach' (Addison-Wesley Publishing Company, 1992)

10 WIRFS-BROCK, R., WILKERSON, B. and WIENER, L.: 'Designing object-oriented software' (Prentice-Hall, 1990)

11 GRAHAM, I.: 'Object-oriented methods' (Addison-Wesley Publishing Company, 1994)

12 SCHACH, S. R.: 'Software engineering with Java' (McGraw-Hill International editions, 1997)

Chapter 3

Static modelling

"(Camelot!) It's only a model"
– Patsy

3.1 Introduction

Previously we have discussed the point that each system must be modelled from several different viewpoints. Each viewpoint represents a different type of model. This chapter examines one of these viewpoints: the static model.

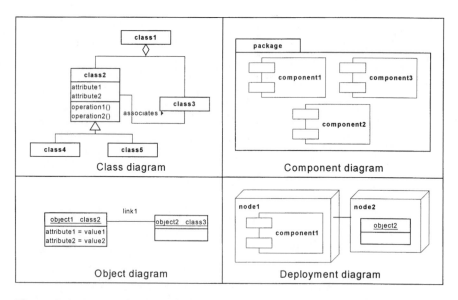

Figure 3.1 *Types of static model*

Figure 3.1 shows the four types of diagram that can be used to realise static models: class diagrams, object diagrams, component diagrams and deployment diagrams.

In order to illustrate the concepts behind static modelling, one of the four static diagrams will be used to show some simple examples. The diagram chosen is the class diagram as this forms the backbone of the Unified Modelling Language (UML). In addition, the static modelling principles that will be shown in this section with reference to class diagrams can be applied when using any type of static model. For more in-depth discussion concerning class diagrams and static modelling, see [1], [2] and [3].

3.1.1   Static modelling using class diagrams

3.1.1.1   Modelling classes and relationships

The class diagram is, arguably, the most widely used diagram in the UML. Class diagrams have a long history and are present in some shape or form in all of the other methodologies that were mentioned previously in this part of the book. The class diagram is also the richest diagram in terms of the amount of syntax available to the modeller. As with all UML diagrams, it is not necessary to use every piece of syntax, as experience has shown that 80 per cent of any modelling task can be achieved by using approximately 20 per cent of class diagram syntax. The simplest (and wisest) approach is to learn the bare minimum syntax and then to learn more as and when circumstances dictate.

3.1.2   Basic modelling

There are two basic elements that make up a class diagram, which are the 'class' and the 'relationship' and, at a very simple level, that is it! Clearly, there are many ways to expand on these basic elements, but providing that they are understood clearly and simply, the rest of the syntax follows on naturally and is very intuitive.

A 'class' represents a type of 'thing' that exists in the real world and, hence, should have a very close connection to reality. Classes are almost always given names that are nouns, as nouns are 'things' and so are classes. This may seem a trivial point, but it can form a very powerful heuristic when assessing and analysing systems as it can be an indicator of whether something may appear as a class on a model.

The second element in a class diagram is a relationship that relates together one or more class. Relationships should have names that form sentences when read together with their associated classes. Remember that the UML is a language and should thus be able to be 'read' as one would read any language. If a diagram is difficult to read, it is a fairly safe bet that it is not a very clear diagram and should perhaps be 'rewritten' so that it can be read more clearly. Reading a good UML diagram should not involve effort or trying in the same way that any sentence should not be difficult to read.

Now that the basics have been covered, it is time to look at example class diagrams. It should be pointed out that from this point, all diagrams that are shown will be UML diagrams. In addition, although some of the diagrams may seem trivial, they are all legitimate and correct diagrams and convey some meaning – which is the whole point of the UML!

Figure 3.2   *Representing classes*

Figure 3.2 shows two very simple classes. Classes are represented by rectangles in the UML and each must have a name, which is written inside the rectangle. In order to understand the diagram, it is important to read the symbols. The diagram here shows that two classes exist: 'Class 1' and 'Class 2'. This is the first UML diagram in the book and, if you can read this, then you are well on the way to understanding the UML.

Figure 3.3   *Representing a relationship*

Figure 3.3 shows how to represent a relationship between two classes. This particular relationship is known as an 'association' and is simply a general type of relationship that relates together one or more class. The association is represented by a line that joins two classes, with the association name written somewhere on the line. This diagram reads: two classes exist: 'Class 1' and 'Class 2' and 'Class 1' associates 'Class 2'.

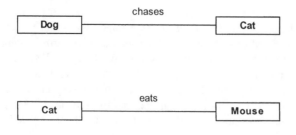

Figure 3.4   *Examples of classes and associations*

Figure 3.4 shows two more examples that are based on real life. The top part of the diagram reads: there are two classes: 'Dog' and 'Cat' where 'Dog' chases 'Cat'. Likewise, the lower part of the diagram reads: there are two classes: 'Cat' and 'Mouse' where 'Cat' eats 'Mouse'.

This illustrates a very important point concerning classes, as classes are abstract and do not actually exist in the real world. There is no such thing as 'Cat', but there do exist many examples of 'Cat'. A class represents a grouping of things that look and behave in the same way, as, at one level, all examples of 'Cat' will have a common set of features and behaviours that may be represented by the class 'Cat'. What this class is really representing is the blueprint of 'Cat', or the essence of 'Cat'.

Another important point illustrated here is that every diagram, no matter how simple, has the potential to contain a degree of ambiguity. Figure 3.4 is actually ambiguous as it could be read in one of two ways, depending on the direction in which the association is read. Take, for example, the top part of the diagram: who is to say that the diagram is to be read 'Dog' chases 'Cat' rather than 'Cat' chases 'Dog', as it is possible for both cases to be true. It should be pointed out that for this particular example the world that is being modelled is a stereotypical world where dogs always chase cats, and not the other way around. Therefore, there is some ambiguity as the diagram must only be read in one direction for it to be true; thus, a mechanism is required to indicate direction.

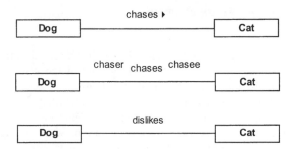

Figure 3.5   *Showing direction*

The simplest way to show direction is to place a direction marker on the association that will dictate which way the line should be read, as shown in the top part of Figure 3.5. The diagram now reads 'Dog' chases 'Cat' and definitely not 'Cat' chases 'Dog' and is thus less ambiguous than Figure 3.4.

The second way to show direction is to define a 'role' on each end of the association, as shown in the middle part of Figure 3.5. In this case, the two roles that have been defined are 'chaser' and 'chasee', which again eliminates the ambiguity that existed in Figure 3.4.

The lower part of Figure 3.5 introduces a new association called 'dislikes'. This time, however, the lack of direction is intentional as both statements of 'Dog' dislikes 'Cat' and 'Cat' dislikes 'Dog' are equally true. Therefore, when no direction is indicated, it is assumed that the association can be read in both directions or, to put it another way, the association is said to be 'bi-directional'.

Another reason why the diagram may be misunderstood is because there is no concept of the number of cats and dogs involved in the chasing of the previous diagrams. Expressing this numbering is known as 'multiplicity', which is illustrated in Figure 3.6.

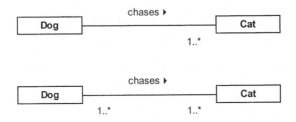

Figure 3.6   *Showing numbers using multiplicity*

The top part of Figure 3.6 shows that each 'Dog' chases one or more 'Cat'. If no number is indicated, as in the case of the 'Dog' end of the association, it is assumed that the number is 'one'. Although the number is one, it does not necessarily indicate that there is only one dog, but rather that the association applies for each dog. The multiplicity at the other end of the 'chases' association states '1..*', which means 'one or more' or somewhere between one and many. Therefore, the association shows that each 'Dog' chases one or more 'Cat'.

The lower part of the diagram shows a case where the multiplicity has been changed, which changes the entire meaning of the model. In this case, the diagram is read as: one or more 'Dog' chases one or more 'Cat'. This could mean that a single dog chases a single cat, a single dog chases any number or a herd of cats, or that an entire pack of dogs is chasing a herd of cats.

Which of the two examples of multiplicity is correct? The answer will depend on the application that is being modelled and it is up to the modeller to decide which is the most accurate of the two.

3.1.2.1   Adding more detail to classes

So far, classes and relationships have been introduced, but the amount of detailed information for each class is very low. After all, it was said that the class 'Cat' represented all cats that looked and behaved in the same way, but it is not defined anywhere how a cat looks or behaves. This section examines how to add this information to a class by using 'attributes' and 'operations'.

Consider again the class of 'Cat', but now think about what general properties the cat possesses. It is very important to limit the number of general properties that are identified to only those that are relevant, as it is very easy to get carried away and over-define the amount of detail for a class.

For this example, suppose that we wish to represent the features 'age', 'weight', 'colour' and 'favourite food' on the class 'Cat'. These features are represented on the class as 'attributes' – one for each feature.

Figure 3.7   *Attributes of the class 'Cat'*

Attributes are written in a box below the class name box. When modelling software, it is possible to add more detail at this point, such as the visibility of the attribute, type, default values, and so forth. As attributes represent features of a class, they are usually represented by nouns and they must also be able to take on different values. For example, 'colour' is a valid attribute, whereas 'red' would not be, as 'red' would represent an actual value of an attribute rather than an attribute itself.

Attributes thus describe how to represent features of a class, or to show what it looks like, but they do not describe how the class behaves, which is represented using 'operations'. Operations show what a class does, rather than what it looks like, and are thus usually represented by verbs. Verbs, as any schoolchild will be able to tell you, represent 'doing' words, and operations describe what a class 'does'. In the case of the class 'Cat' we have identified three things that the cat does, which are 'eat', 'sleep' and 'run'.

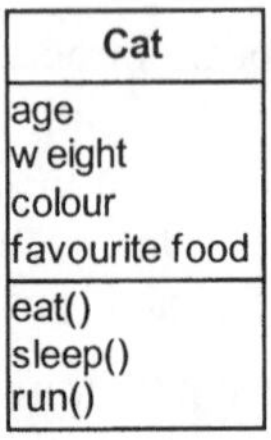

Cat
age w eight colour favourite food
eat() sleep() run()

Figure 3.8 *Operations of the class 'Cat'*

Operations are represented in the UML by adding another rectangle below the attribute rectangle and writing the operation names within it. When modelling software, operations may be defined further by adding extra detail – for example, arguments, return values, visibility, and so forth.

The class 'Cat' is now fully defined for our purposes and the same exercise may be carried out on any other classes in the diagram in order to fully populate the model. It should also be pointed out that the classes may be left at a high level with no attributes or operations. As with everything in the UML, only use as much detail as is necessary, rather than as much as is possible.

3.1.2.2 Adding more detail to relationships

Section 3.1.2.1 showed how to add more detail to classes, while this section shows how to add more detail to relationships, by defining some special types that are commonly encountered in modelling. There are four types of relationship that will be discussed here: 'association', 'aggregation', 'specialisation' and 'instantiation'. Many types of relationship exist, but these four represent the majority of the most common uses of relationships.

Associations have already been introduced and shown to be a very general type of relationship that relate together one or more class. Therefore, they will not be discussed in any further detail, so the next three sections cover each of the other special types of relationship.

3.1.2.3 Aggregation and composition

The second type of relationship is a special type of association that allows assemblies and structures to be modelled and is known as 'aggregation'.

Figure 3.9 provides an example of aggregation. Aggregation is shown graphically in the UML by a diamond or rhombus shape and when reading the diagram is read by saying 'is made up of'. Starting from the top of the diagram, the model is read as: 'Cat' wears 'Collar'. The direction is indicated with the small arrow and there is a one-on-one relationship between the two classes. The multiplicity here is implied to be one to one as there is no indication. This makes sense as it is very unlikely that one cat would wear one or more collar and almost impossible to imagine that more than one cat would wear a single collar!

The 'Collar' is made up of (the aggregation symbol) a 'Bell', a 'Belt' and a 'Buckle'. The 'Bell' is on the 'Belt' and the 'Buckle' is on the 'Belt'.

The 'Bell' is made up of (the aggregation symbol) a 'Clasp', a 'Donger' and a 'Sphere'.

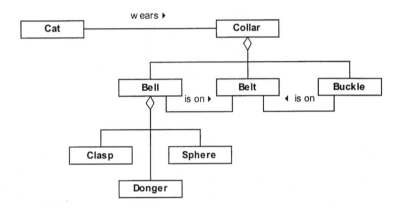

Figure 3.9 *Example of aggregation*

This is the basic structure of the bell and allows levels of abstraction of detail to be shown on a model.

There is also a second special type of association that shows an aggregation-style style of relationship, known as 'composition'. The difference between composition and aggregation is subtle but very important and can convey much meaning. The simplest way to show this difference is to consider an example, as shown in Figure 3.10.

The model in Figure 3.10 represents the static structure of the types of weapon that may be used in the sport of fencing (as opposed to putting up fences or selling stolen goods). From the model, there are three types of 'Weapon': 'Foil', 'Epée' and 'Sabre' (this is a 'type of', or 'specialisation' relationship that will be discussed in a subsequent section). Each weapon is made up of a 'Handle', a 'Pommel', a 'Blade' and a 'Guard'. The 'Blade' is made up of a 'Forté', a 'Foible' and a 'Tang'. There is a clear visual difference here as the aggregation symbol under 'Blade' is filled in, rather than empty as in the case of the aggregation under 'Weapon'. The semantic difference, however, is that   an

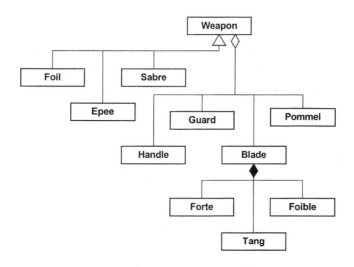

Figure 3.10 *Example of the difference between composition and aggregation*

aggregation is made up of component parts that may exist in their own right. It is possible to buy or make any of the components under the class 'Weapon' as they are assembled into the completed class 'Weapon'. The class 'Blade', however, has three components that cannot exist independently of the class 'Blade'. This is because a fencing blade is a single piece of steel that is composed of three distinct sections. For example, there is no such thing as a 'Foible' as it is an inherent part of the 'Blade' rather than being an independent part in its own right.

3.1.2.4 *Specialisation and generalisation*

The third special type of relationship is known as 'specialisation and generalisation', depending on which way the relationship is read. 'Specialisation' refers to the case when a class is being made more special or is being refined in some way. Specialisation may be read as 'has types' whenever its symbol, a small triangle, is encountered on a model. If the relationship is read the other way around, then the triangle symbol is read as 'is a type of', which is a generalisation. Therefore, read one way the class becomes more special (specialisation) and read the other way, the class becomes more general (generalisation).

Specialisation is used to show 'child' classes, sometimes referred to as subclasses, of a 'parent' class. Child classes inherit their appearance and behaviour from their parent classes, but will be different in some way in order to make them special. In UML terms, this means that a child class will inherit any attributes and operations that its parent class has, but may have additional attributes or operations that make the child class special.

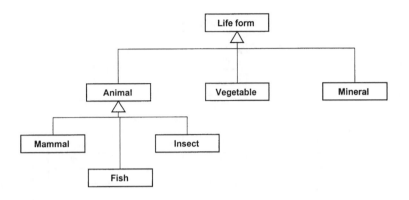

Figure 3.11 *Life form hierarchy*

As an example of this, consider Figure 3.11, which shows different types of life known to man.

The top class is called 'Life form' and has three child classes: 'Animal', 'Vegetable' and 'Mineral', which makes 'Life form' the parent class. Going down one level, it can be seen that 'Animal' has three child classes: 'Mammal', 'Fish' and 'Insect'. Notice now how 'Animal' is the parent class to its three child classes, while still being a child class of 'Life form'.

The diagram may be read in two ways:

- From the bottom up: 'Mammal', 'Fish' and 'Insect' are types of 'Animal'. Both 'Animal' and 'Vegetable' are types of 'Life form'.

- From the top down: 'Life form' has two types: 'Animal' and 'Vegetable'. The class 'Animal' has three types: 'Mammal', 'Fish' and 'Insect'.

This has explained how to read the diagrams, but has not covered one of the most important concepts associated with generalisation and specialisation, which is the concept of inheritance, best illustrated by considering the following diagram.

Figure 3.12 shows an expanded version of Figure 3.11 by adding some attributes and operations to the classes. It can be seen that the class 'Animal' has three identifiable attributes: 'age', 'gender' and 'number of legs'. These attributes will apply to all types of animal and will therefore be inherited by their child classes. That is to say that any child classes will automatically have the same three attributes. This inheritance also applies to the behaviour of the parent class; therefore, if it were the case where some operations had been defined for the class

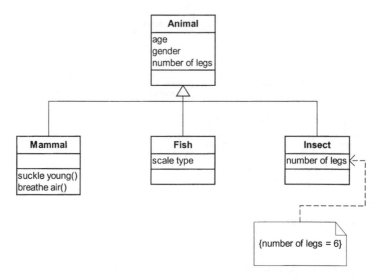

Figure 3.12 *Example of inheritance*

'Animal', these would also be inherited by each of the child classes. What makes the child class different or special and therefore an independent class in its own right, is the addition of extra attributes and operations or constraints on existing attributes. Let us consider an example of each one of these cases:

- The class 'Mammal' has inherited the three attributes from its parent class. Inherited properties (attributes and operations) are not usually shown on child classes. In addition, it has had an extra two operations identified, which the parent class does not possess: 'breathe air' and 'suckle young'. This is behaviour that all mammals possess, whereas not all animals will, or to put it another way, why mammals are special compared to animals generally.

- The class 'fish' has inherited the three attributes from its parent class (which are not shown), but has had an extra attribute added that its parent class will not possess: 'scale type'. This makes the child class more specialised than the parent class.

- The class 'insect' has no extra attributes or operations but has a constraint on one of its attribute values. The attribute 'number of legs' is always equal to six, as this is in the nature of insects. The same could be applied to 'mammal' to some extent as 'number of legs' would always be between zero (whales and dolphins) and four, while 'number of legs' for 'fish' would always be zero. Such a limitation is known in the UML as a 'constraint', which is one of the standard extension mechanisms for the UML and which is covered in some detail in Chapter 8.

From a modelling point of view, it may be argued that the attribute 'number of legs' should not be present in the class 'animal' as it is not applicable to fish. This is fine and there is nothing inherently wrong with either model except to say that it is important to pick the most suitable model for the application at hand. Remember that there are many correct solutions to any problem, and thus people's interpretation of information may differ. By the same token, it would also be possible to define two child classes of 'animal' called 'male' and 'female', which would do away with the need for the attribute 'gender' as shown in Figure 3.13.

The model in Figure 3.13 shows another approach to modelling the gender of an animal. Which approach is the better of the two, the one shown in Figure 3.12 or the one shown in Figure 3.13? Again, it is necessary to pick the most appropriate visual representation of the information and one that you, as the modeller, are comfortable with.

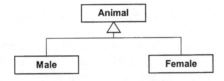

Figure 3.13 *Another way to model gender*

As a final example, let us revisit our old friend the cat who is undeniably a life form, an animal and also a mammal. This gives three layers of inheritance, so that the class 'Cat' may now be represented as in Figure 3.14.

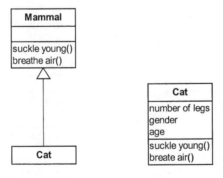

Figure 3.14 *A cat's inheritance*

The model in Figure 3.14 shows an example of the cat that is now applied to the life form hierarchy. Here, it can be seen that 'Cat' is a type of 'Mammal' and therefore inherits all its features and behaviour from its parent class. The class 'Cat' shown on the right-hand side of the diagram shows all the inherited attributes and operations from its parent classes. Note the inclusion of the three attributes that were inherited from 'Animal', which, although the class is not included on this particular diagram, still influences the child classes, no matter how far removed.

3.1.2.5 Instantiation

The final type of relationship that will be defined in this book is known as 'instantiation', which is a way of showing real-life examples of a particular class. These examples are known as 'instances' and an instance of a class is also known as an 'object'.

Strictly speaking, the instantiation relationship is actually a special type of relationship that is known as a 'dependency'. A dependency, as the name implies, relates two classes together but indicates that if one class changes in any way, so will its dependent class. A dependency relationship is shown graphically by using a dashed directed line, which indicates the direction of the dependency. Several other types of dependency exist, but they will not be covered in this book. Clearly, an instantiation will be a dependency relationship as an object is entirely dependent on its class to define its features and behaviour.

Instances are represented in the UML by underlining the class name and showing the instance name next to the class name, separated by a colon, as shown in Figure 3.15. It is also possible to use shorthand to represent some instances.

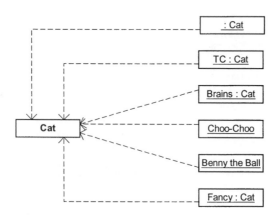

Figure 3.15 *Example of instantiation*

From the model, it can be seen that the class 'Cat' has six instances: 'TC', 'Brains', 'Choo-Choo', 'Benny-the-Ball', 'Fancy' and an extra instance with no name. These instances are represented in three ways:

- The full representation of the instance: instance name, a colon and the class name all underlined. This is shown by "<u>TC:Cat</u>" and "<u>Brains:Cat</u>".

- An abbreviated instance, where only the instance name is shown, as in 'Choo-Choo' and 'Benny-the-Ball'. This leaves room for ambiguity, however, as it is assumed that whoever is reading the diagram has an assumed knowledge concerning the classes of these instances. This is useful if all the instances on a particular model are of the same class and serves as a type of shorthand.

- The final example of the class 'Cat' is an anonymous instance, which means that although there is definitely a real-life example of the class 'Cat', there is no specific name to differentiate it from other examples of the class 'Cat'.

One point that should be borne in mind here is that the instance name is a unique identifier, rather than the actual name of the cat. There should only be one instance for each different identifier, as these identifiers need to be unique. What would happen, however, if an instance were named after the actual cat's name (TC for example) and the other cat appeared with the same name? It is impossible to have two objects, or instances, that are both called TC, so how can this be resolved? The answer would be to add another attribute called 'name' and then fill it in for each object, making the word chosen for the unique identifier irrelevant.

Instantiation completes the list of types of relationship that will be introduced in this book, but remember that there are many other types that are used for specific applications or under particular conditions. Reference to these can be found in any other UML book, a number of which are provided at the end of this chapter.

3.1.3 Other static diagrams

This chapter is intended to introduce static modelling as a concept, but has so far only considered one of the four static diagrams: the class diagram. However, the concepts introduced here, that of things and the relationships between them, applies to each of the static diagrams. Therefore, if you can understand the concepts used here, then it is simply a matter of learning some syntax and some example applications for the other types of diagram. Indeed, this is achieved in Chapter 5 where each of the nine UML diagrams is described in terms of its syntax and its typical uses. The important thing to realise about all four types of static diagram is that they all represent the 'what' of a system. These other four diagrams are the object diagram, the component diagram and the deployment diagram.

3.1.4   Conclusion

In order to conclude this section and in order to see if static modelling really works, a summary of this section is presented in the following class diagram:

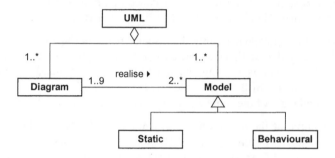

Figure 3.16 *Summary of UML models and their associated diagrams*

Figure 3.16 shows a summary of the UML models and how they may be realised using the various types of diagram. The 'UML' is made up of two types of 'Model': 'Static' and 'Behavioural'. Each 'Model' may be realised by between one and nine 'Diagram'.

The actual types of diagram are not shown here, but are shown in the following model, which introduces a package that groups together the various types of static model.

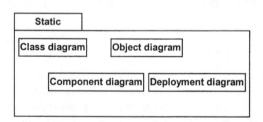

Figure 3.17 *Package of the types of static model*

Figure 3.17 shows the various types of static diagram. It can be seen from the model that the package 'Static' (a simple grouping) contains 'Class diagram', 'Object diagram', 'Deployment diagram' and 'Component diagram'.

The actual components that made up the class diagram can also be modelled using a class diagram. This is a concept known as the 'meta-model', which will be used extensively in Chapter 5.

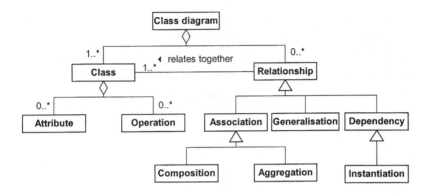

Figure 3.18 *Meta-model for a class diagram*

Figure 3.18 shows the UML meta-model for a class diagram. It can be seen that a 'Class diagram' is made up of one or more 'Class' and zero or more 'Relationship'. Each 'Class' is made up of zero or more 'Attribute' and zero or more 'Operation'. A 'Relationship' describes the relationship between one or more 'Class' and has several types: 'Association' (which has types 'Aggregation' and 'Composition'), 'Instantiation' and 'Dependency' (which has a type 'Instantiation').

This meta-model concept may be extended to include all the other types of static diagram and, indeed, this is exactly the approach adopted in Chapter 5 to introduce all the other UML diagram types.

3.2 Further discussion

1. Create a new class based on an everyday object and model its components. Should the relationships be aggregation or composition?

2. Define a new type of 'Mammal' and represent it as a specialisation. Define some new attributes and operations that will make it special when compared to its parent class.

3. Define a completely new hierarchy of animals, but rather than defining types based on genetics, create new types based on the way that they move around. For example, land-based, air-based and water-based. How do the types of animal already defined fit in here? Indeed, do they still fit in?

4. Define some new instances of the new types of 'Mammal'. Show the attributes on each object and fill in the values of each attribute.

3.3   References

1   BOOCH, G., RUMBAUGH, J. and JACOBSON, I.: 'The unified modelling language user guide' (Addison Wesley, 1998)

2   RUMBAUGH, J., JACOBSON, I. and BOOCH, G.: 'The unified modelling language reference manual' (Addison Wesley, 1998)

3   STEVENS, P. and POOLEY, R.: 'Using UML' (Addison Wesley, 1999)

Chapter 4

Behavioural modelling

"Oh, behave"
– Austin Powers

4.1 Introduction

Chapter 3 introduced static modelling, one of the two basic types of modelling using the Unified Modelling Language (UML), by choosing one type of static diagram and using it to explain basic principles that may then be applied to any sort of static modelling. This chapter takes the same approach, but with behavioural modelling.

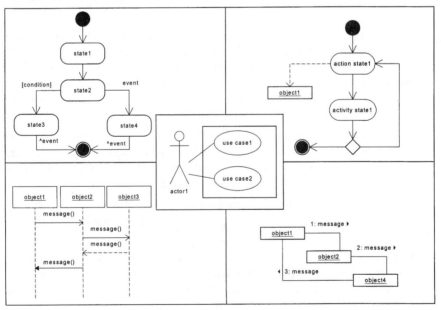

Figure 4.1   *The five diagrams for behavioural models*

Behavioural models may be realised using five types of UML diagram, as shown in Figure 4.1.

Figure 4.1 shows the four types of behavioural UML diagram, which are: use case diagrams, statecharts, activity diagrams and interaction diagrams, of which there are two types: collaboration and sequence diagrams.

It was stated in Chapter 3 that static modelling defined the 'what' of a system: what it looks like, what it does and what the relationships are. If static modelling tells us 'what', then behavioural modelling tells us 'how'. This 'how' is described by modelling interactions within a system. These interactions may be modelled at many levels of abstraction; different types of behavioural diagram allow the system to be modelled at different levels of abstraction. These levels of abstraction may be categorised as follows:

- Interactions may be modelled between objects or between subsystems. Such models are realised using the two types of interaction diagram: collaboration or sequence diagram.

- Interactions may also be modelled at a slightly lower level of abstraction by considering the interactions within a class or its associated objects. Modelling the behaviour over the lifetime of a class, or an object, is achieved by using statechart diagrams. Statecharts are concerned with one, and only one, class and its objects, although it is possible to create more than one statechart per class.

- The lowest level of abstraction for modelling interactions is concerned with the interactions within an operation or, in computer terms, at the algorithmic level. Interactions at the algorithmic level are modelled using activity diagrams. Activity diagrams may also be used to model interactions at a higher level of abstraction, which will be discussed in greater detail in Chapter 5.

- The remaining level of abstraction is at the highest possible level, which is the context level of the project, where high-level functionality of the system and its interaction with the outside world is modelled. These high-level interactions are realised using use case diagrams.

The diagram chosen to illustrate behavioural modelling is the statechart diagram. There are a number of reasons for choosing this behavioural diagram over the other four.

- Statecharts are one of the most widely used diagrams in previous modelling languages and, as such, people tend to have seen something similar to statecharts, if not an actual statechart. This makes the process of learning the syntax easier.

- Statecharts have a very strong relationship with class diagrams, which were discussed in Chapter 3.

It will be seen later in this chapter that behavioural modelling using statecharts will lead directly into other types of behavioural modelling diagrams. For more in-depth discussion concerning statechart diagrams and behavioural modelling, see [1], [2] and [3].

4.1.1 Behavioural modelling using statecharts

4.1.1.1 Introduction
Statecharts are used to model behaviour over the lifetime of a class, or to put it another way, they describe the behaviour of objects. Both of these statements are true and are used interchangeably in many books. Remember that objects are instances of classes and thus use the class as a template for the creation of its objects, which includes any behaviour.

The most obvious question to put forward at this point is 'does every class need to have an associated statechart?' The simple answer is 'no', as only classes that exhibit some form of behaviour can possibly have their behaviour defined. Some classes will not exhibit any sort of behaviour, such as data structures or database structures. The simple way to spot whether a class exhibits any behaviour is to see whether it has any operations; if a class has operations, it does something, if it does not have any operations, it does nothing. If a class does nothing, it is impossible to model the behaviour of it. Therefore, a simple rule of thumb is that any class that has one or more operation must have an associated statechart.

4.1.1.2 Basic modelling
The basic modelling elements in a statechart are states, transitions and events. States describe what is happening within a system at any given point in time, transitions show how to change between such states and events dictate which messages are passed on these transitions. Each of these elements will now be looked at in more detail, starting with the state, an example of which is shown in Figure 4.2.

Figure 4.2 *A UML representation of a state*

Figure 4.2 shows a very simple state, which is shown in the UML by a box with rounded corners. This particular state has the name 'State 1' and this diagram should be read as: 'there is a single state, called "State 1"'. This shows what a state looks like, but what exactly is a state? The following three points discuss the basics of a state:

- A state may describe situations in which the system satisfies a particular condition, in terms of its attribute values or events that have occurred. This may, for example, be 'loaded' or 'saved', so that it gives an indication as to something that has already happened. States that satisfy a particular condition tend to be used when an action-based approach is taken to creating statecharts. This will be discussed in more detail in due course.

- A state may describe a situation in which the system performs a particular activity or set of actions, or to put it another way, is actually doing something. States are assumed to take a finite amount of time, whereas transitions are assumed to take no time. There are two things that can be happening during such a state: an activity (which takes a finite amount of time to execute and is directly equivalent to an operation from a statechart's parent class) and an action (which is assumed to take zero time and can be placed either inside a state or on a transition). Activities are assumed to take time.

- A state may also describe a situation in which a system does nothing or is waiting for an event to occur. This is often the case with event-driven systems, such as windows-style software where, in fact, most of the time the system is sat idle and is waiting for an event to occur.

- There are different types of state in the UML; however, the states used in statecharts are known as normal states. This implies that complexity may exist within a state, such as: a number of things that may happen, certain actions that may take place on the entry or exit of the state, and so forth.

In order for the object to move from one state to another, a transition must be crossed. In order to cross a transition, some sort of event must occur. Figure 4.3 shows a simple example of how states and transitions are represented using the UML.

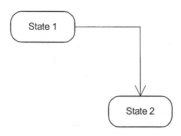

Figure 4.3   *States and transitions*

From the model in Figure 4.3 it can be seen that two states exist: 'State 1' and 'State 2', represented by rounded boxes. There is a single transition that goes from 'State 1' to 'State 2', which is represented by a directed line that shows the direction of the transition. These transitions are unidirectional and, in the event of another transition being required going in the other direction, an entirely new transition is required – the original transition cannot be made bi-directional.

In order to cross a transition, which will make the object exit one state and enter another, an event must occur. This event may be something simple, such as the termination of an activity in a state (the state has finished what it is doing), then it leaves its present state, or may be more complex and involve receiving messages from another object in another part of the system. Event names are written on the transition lines.

Statecharts will now be taken one step further by considering a simple example.

4.1.2 Behavioural modelling – a simple example

4.1.2.1 Introduction

In order to illustrate the use of statecharts, a simple example will be used that will not only show all basic concepts associated with statecharts, but will also provide a basis for further modelling and provide a consistent example that will be used throughout this section.

The simple example chosen is that of a game of chess as it is a game with which most people are at least vaguely familiar, and thus most people will have some understanding of what the game is all about. In addition, chess is, on one level, very simple, which means that it can be modelled very simply, yet, on the other hand, it possesses a great deal of hidden complexity that should emerge as the example is taken further.

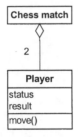

Figure 4.4 *A simple class diagram for a game of chess*

Figure 4.4 shows a class diagram that represents, at a very simple level, a game of chess. From the model, a 'Chess game' is made up of two 'Player'. Each 'Player' has attributes – 'result' and 'status' – and a single operation 'move'. The attribute 'result' reflects the result of the game and may have values: 'this player win', other player win', 'draw' or 'undecided'. The attribute 'status' reflects the current status of the game and may take the values 'checkmate', 'stalemate' and 'game in progress'.

4.1.2.2   Simple behaviour

It was stated previously that any class that exhibits behaviour must have an associated statechart. Applying this rule reveals that, of the two classes present in the class diagram, only the class 'Player' needs to have a statechart. A very simple statechart for a game of chess is shown in Figure 4.5, by defining the behaviour of the class 'Player'.

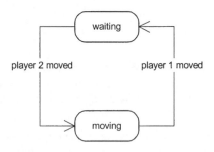

Figure 4.5   *A simple statechart for a game of chess*

From the model, the object may be in one of two states: 'waiting' or 'moving'. In order to cross from 'waiting' to 'moving', the event 'player 2 moved' must have occurred. In order to cross from 'moving' to 'waiting', the event 'player 1 moved' must have occurred.

At a very simple level, this is 'how' a game of chess is played, by modelling the behaviour of each player. However, the model is by no means complete as the chess game described here has no beginning or end and will thus go on forever. Despite the fact that chess games may seem to go on forever, there is an actual distinct start and end point for the game. This is modelled in the UML by introducing 'start states' and 'end states'.

The next step, therefore, is to add more detail to this statechart. It is interesting to note that this is actually how a statechart (or any other UML diagram, for that matter) is created. The diagram almost always starts off as a simple collection of states and then evolves over time and, as more detail is added, so the model starts to get closer and closer to the reality that it is intended to represent.

4.1.2.3 Adding more detail

The next step, as stated in Section 4.1.2.2, is to add a beginning and an end for the statechart, using start and end states. A start state describes what has happened before the object is created and is shown visually by a filled-in circle. An end state, by comparison, shows the state of the object once the object has been destroyed and is represented visually by two concentric circles, the smaller of which is filled in.

Start and end states are treated just like other states in that they require transitions to take the object into another state, which will require appropriate events. The event would typically be creation- and destruction-type events that are responsible for the birth and subsequent demise of the object.

Figure 4.6 shows the expanded statechart that has start and end states along with appropriate events.

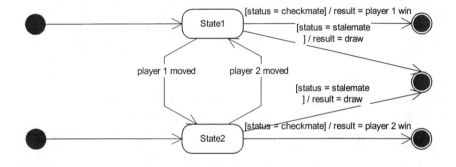

Figure 4.6   *Expanded statechart showing start and end states*

It can be seen from the model that there are two possible start states, depending on which player goes first, and three different end states, depending on which player, if either, wins, or whether the game is a draw. The model is now becoming more realistic; its connection to reality is getting closer, but there is still room for ambiguity. We know from the class diagram that the class 'Player' does one thing, 'move', but we do not know in which state is the statechart that this operation is executed. As it happens, it is fairly obvious which state the operation, occurs in: 'moving'. Operations on a statechart may appear as either 'activities' or 'actions', which will be discussed in more detail in due course. This may now be shown by adding the activity to its appropriate state by writing 'do/' in the state box and then adding the activity (operation from the class diagram) name, which is shown in Figure 4.7.

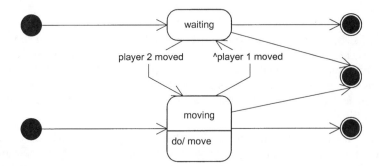

Figure 4.7   *Expanded statechart showing activity*

The model is now getting even closer to reality; in fact, the model is evolving, which they always do! It is almost impossible to get a model right the first time as they are living entities and will continue to evolve for as long as they exist. However, there is yet another problem with the statechart, as, although it seems to work well for any situation in a chess game, it is impossible for the game to run! To illustrate this, consider what happens when we begin a game of chess.

4.1.2.4   Ensuring consistency

The first thing that will happen is that two instances of the class "Player" need to be created so that we have the correct number of players. The behaviour of each player is described by the statechart for the class "Player". For arguments sake, we shall name the two players "Player 1" and "Player 2" and see if the statechart will hold up to the full game of chess. Figure 4.8 shows two, identical statecharts, one for each object, that are positioned side-by-side to make comparisons easier.

In order to begin a game of chess, an instance of the class 'Chess match' would be created, which would in turn create two instances of the class 'Player'. In this example, the object names 'Player 1' and 'Player 2' have been chosen. Let us now imagine that a game of chess has been started and that 'Player 1' is to begin. The event that will occur is 'player 1 start', which is present on both statecharts. However, this will put both players straight into the 'moving' state, which will make the game of chess impossible to play, despite being slightly more entertaining. This is because the events were named with reference to one player, rather than being generic so that they are applicable to any player. In order to make the game work, it is necessary to rename the events so that they are player-independent.

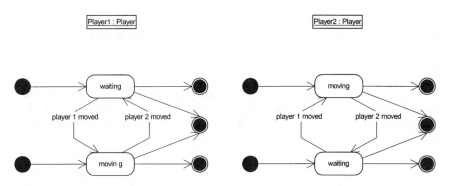

Figure 4.8 *Side-by-side comparison of two statecharts*

This serves to illustrate a very important point that relates back to the section about modelling: the different levels of abstraction of the same model. The chess game was modelled only at the object level in terms of its behaviour, which, it is entirely possible, would have resulted in the class being implemented and even tested successfully if treated in isolation and under test conditions. It would only be at the final system test level that this error would have come to light. It would be useful, therefore, if it were possible to model behaviour at a higher level, where the interactions between objects could be shown, as in Figure 4.9.

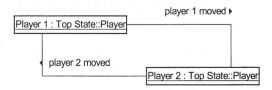

Figure 4.9 *Wouldn't it be nice model*

Figure 4.9 shows the two objects from the original class diagram, but this time the events from the statecharts have been shown. It is clear from this model that there would be a problem with the event names at a higher level of abstraction, which could lead to the problem being sorted out far earlier than the previous case. It would have been nice if we had drawn this model as part of our behavioural modelling of the game of chess. Luckily, such a model does exist and is known as a collaboration diagram, which will be covered in detail in Chapter 5.

4.1.3 Solving the inconsistency

There are many ways to solve the inconsistency problems that were highlighted in the previous section, two of which are presented here. The first solution is to make the generic statechart correct, while the second is to change the class diagram to make the statecharts correct.

4.1.3.1 Changing the statechart

The first solution to the inconsistency problem that is presented here is to make the statechart more generic, so that the events now match up. The statechart for this solution is shown in Figure 4.10.

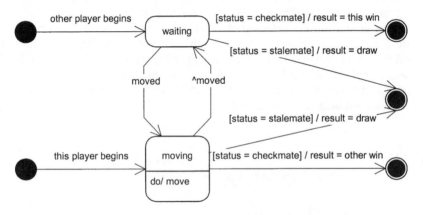

Figure 4.10 *New statechart with correct event names*

Figure 4.10 represents a correct solution to the chess model. Note that in this model the event names have been changed to make them more generic. The first events are the event on the transitions from the two start states. Note that this time the names have been changed so that they are no longer specific to either player one or player two. The names are now relative to each instance, rather than specific, so the event names have changed from 'Player 1 begins' to 'this player begins' and from 'player 2 begins' to 'other player begins'.

The other event names that have been changed are on the transitions between the two normal states. Previously, the names were object-specific, being as they were, 'Player 1 moved' and 'Player 2 moved'. These names have now changed to a more-generic 'moved' event, which will apply equally to both objects.

This is by no means the only solution to the problem and another possible solution is presented in Section 4.1.4, where the class diagram is changed to make the statechart correct, rather than changing the statechart.

4.1.3.2 Changing the class diagram

The second solution to the consistency problem is to change the class diagram rather than the statechart, as shown in Figure 4.11.

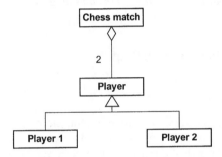

Figure 4.11 *A modified class diagram*

Figure 4.11 shows a modified class diagram in which two new subclasses of 'Player' have been added. This would mean that, rather than the class 'Player' being instantiated, one instance of each class 'Player 1' and 'Player 2' would be created. This has implications on the statecharts as the class diagram shown here would require a statechart for both 'Player 1' and 'Player 2', rather than a single statechart for 'Player'. This would also mean that the initial statechart shown in Figure 4.7 would now be correct for 'Player 1', but that a new statechart would have to be created for the class 'Player 2'.

Taking this idea a step further, it is also possible to make the two subclasses more specific as, in the game of chess, one player always controls white pieces and the other player only controls black pieces. This would have an impact on the class diagram again, as each subclass could now be named according to its colour. This is shown in Figure 4.12.

Figure 4.12 shows a class diagram where the classes have been named according to colour, rather than simply 'Player 1' and 'Player 2'. This has even more effect on the statechart as, in the standard rules of chess, the white player always moves first.

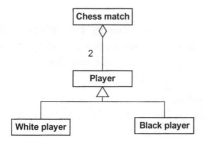

Figure 4.12 *Further modification of the chess class diagram*

4.2   Styles of behavioural modelling

4.2.1   Actions and activities

The statecharts presented so far have had an emphasis on the activities in the system where the main state in the statechart had an explicit activity associated with it. This is slightly different from another approach that is often used in many texts, which is referred to here as an action-based approach. In order to appreciate the difference between the action-based and activity-based approaches, it is important to distinguish between actions and activities.

- An activity describes an ongoing, non-atomic execution within a system. In simple terms, this means that an activity takes time and can be interrupted. An activity is also directly related to the operations on a class.

- An action is an atomic execution within the system that result in either a change in state of the return of a value. In simple terms, this means that an action takes no time and cannot be interrupted.

Activities may be differentiated from actions as they will always appear inside a state (as what is known as an 'internal transition') and will use the keyword 'do/' as a prefix. Any other keyword used as a prefix (for example, 'entry/', 'exit' etc.) is assumed to be an action.

These two types of execution may change the way that a statechart is created and the type of names that are chosen to define its statechart.

4.2.1.1   Activity-based statecharts

Activity-based statecharts are created with an emphasis on the activities within states. Activities must be directly related back to the class that the statechart is describing and must provide the basis for a very simple, yet effective, consistency check between the class diagram and statechart. As they take time to execute, activities may only be present on a statechart within a state, as states are the only element on the statechart that take time – transitions are always assumed to take zero time.

The state names, when using the activity-based approach, tend to be verbs that describe what is going on during the state and thus tend to end with 'ing', such as the two states seen so far in this chapter: 'moving' and 'waiting'.

4.2.1.2 Action-based statecharts

Action-based statecharts are created with an emphasis on actions rather than activities. Actions are often not related directly back to classes, although they should be, in order to ensure that consistency checks are performed between the two types of diagram. Actions are assumed to take zero time to execute and thus may be present either inside states or on transitions – transitions also take zero time. With an action-based approach, the actions tend to be written on transitions rather than inside states, which leads to a number of empty states. These states have names that reflect the values of the system at that point in time.

The simplest way to demonstrate the difference is to create an action-based statechart for the chess example and then to compare the two different styles of statechart.

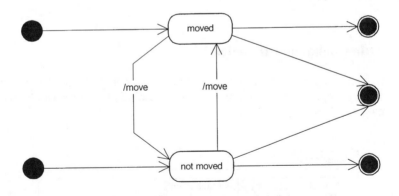

Figure 4.13 *Action-based statechart*

Figure 4.13 shows an action-based statechart where the emphasis is on the action '/move' rather than on activities. Notice that the state names here reflect what has happened in the system, implying that the system is doing nothing while waiting for events.

4.2.2   Comparing the approaches

Each of the two approaches shown here are valid and useful; however, there are a number of observations that can be made concerning the use of both approaches:

- The activity-based approach may lead to a more rigorous approach to modelling as the consistency checks with its associated class are far more obvious.

- The activity-based approach makes use of executions that take time and may be interrupted for whatever reason.

- The action-based approach is more applicable to event-based systems such as windows-based software, where for much of the time the system is idle while waiting for things to happen.

- Although neither approach is particularly suitable for real-time systems, the action-based approach is less suitable then the activity-based approach, as the basic assumption of executions taking zero time goes against much real-time thinking.

- Action-based diagrams tend to be less complex than activity-based, as they tend to have fewer states.

So which approach is the better of the two? The answer is whichever is most appropriate for the application at hand.

4.3   Other behavioural models

This chapter has concentrated mainly on statecharts, but the principles that apply to statecharts apply to the other types of behavioural model. These can be seen in Figure 4.14.

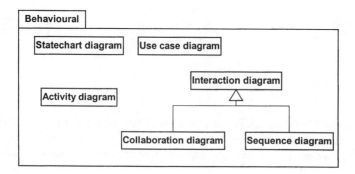

Figure 4.14 *Types of behavioural diagram*

If you can have a good grasp of the models shown here and, just as importantly, understand how they must be checked to ensure consistency between different diagrams (and models), the other diagrams are relatively straightforward.

- Statecharts are used to describe the behaviour of a class and, hence, its associated objects. They represent a medium level of abstraction.

- Activity diagrams emphasise the actions within a system and are generally used at a very low level, or algorithmic level, of abstraction. Activity diagrams may also be used at a medium level of abstraction, but with an emphasis on actions within a system, rather than states. The main difference between an activity diagram and a statechart is that a statechart contains normal states (that may be complex) whereas an activity diagram will only contain 'activity states' or 'action states'. This is elaborated upon in Chapter 5.

- Both types of interaction diagram – collaboration diagrams and sequence diagrams – model interactions at a high level of abstraction and can be used, among other purposes, to ensure that events passed between statecharts are consistent.

- Use case diagrams represent the highest level of abstraction and show the interactions between the system and its external actors. These are strongly related to both types of interaction diagram, which may be used to model scenarios as instances of use cases.

All of the points raised above are examined in Chapter 5, where each diagram is looked at in turn; however, one point that should emerge here is that there is a relationship between each type of behavioural diagram and they each share a number of common features.

4.4 Conclusions

This chapter has introduced behavioural modelling. Behavioural modelling shows the 'how' of a system, whereas static modelling shows the 'what'. Behavioural modelling was illustrated by looking at a single diagram – the statechart – where a simple example was taken and described in some detail.

It was shown that statecharts describe the behaviour within a class, but that modelling at one level of abstraction only could lead to mistakes, despite the fact that the diagram itself may appear to be correct. The solution to this problem was to model the system at more than one level of abstraction by using other behavioural diagrams.

4.5 Further discussion

1. Take the chess example and add a new operation called 'think' to its class diagram, shown in Figure 4.4. The idea behind this is that the 'move' operation will represent the actual physical moving of the chess piece, whereas the 'think' operation represents the planning and strategy involved before the piece is actually moved. Modify the statechart in Figure 4.5 to take into account the new operation.

2. Create two statecharts for the classes 'White player' and 'Black player' shown in Figure 4.12, bearing in mind the rule that states that white always moves first. What effect does this have on the statecharts, particularly with regard to the start states?

3. Apply question 1 to question 2, so that now each of the two diagrams must be modified to take into account the 'think' operation. Is this more convenient that having a single generic statechart for 'Player', or is it more effort?

4. Taking the results of question 1, which, if any, of the two operations may be represented using actions rather than activities? Why is this?

5. Imagine that you are creating a chess computer game and create two new classes of 'Player' called 'Human player' and 'Computer player'. Can a single statechart still be used by both types of player, or do they need individual statecharts?

6. Would it have been beneficial to model the game at the higher level (the collaboration diagram from Figure 4.9) before modelling at the statechart level? Revisit this point after you have read Chapter 5.

4.6 References

1   BOOCH, G., RUMBAUGH, J. and JACOBSON, I.: 'The unified modelling language user guide' (Addison Wesley, 1998)

2   RUMBAUGH, J., JACOBSON, I. and BOOCH, G.: 'The unified modelling language reference manual' (Addison Wesley, 1998)

3   STEVENS, P. and POOLEY, R.: 'Using UML' (Addison Wesley, 1999)

The UML diagrams

try not. Do. Or do not. There is no try
– Yoda

5.1 Introduction

This chapter introduces the nine types of diagram that may be used in the Unified Modelling Language (UML). The information in this chapter is kept, very deliberately, at a very high level. There are several reasons for this:

- The main focus of this book is to provide practical guidelines and examples of how to use the UML efficiently and successfully. It is, therefore, a deliberate move to show only a small subset of the UML language in this chapter. In reality, it is possible to model 80 per cent of any problem with less than 20 per cent of the UML language.

- Experience has shown that the most efficient way to learn, and hence master, the UML is to learn just enough to get going and then to simply try out some models of your own. This chapter aims to give just enough information to allow you to get your hands dirty.

- There are a great many UML books that have been published that are devoted entirely to the UML and its very rich syntax, see references [1–14]. This book should be used in conjunction with such books – for example, with the UML reference manual [1].

This chapter also introduces the concept of the UML meta-model. The UML meta-model is a UML model of the UML, or to put it another way, the UML may be completely modelled using the UML. The UML meta-model presented here is a subset of the whole model. Readers are encouraged to use these models as a starting point and to add to them as their knowledge and experience of the UML increase. It should be pointed out that the meta-models shown here are simple representations of the actual UML meta-model, which can be found in the UML standard. The aim, as with all UML models, is to aid communication; therefore, the models here are very much simplified versions of the actual model. It is also very important to understand what the UML meta-model is when using some of the advanced UML concepts, such as stereotypes, where the UML meta-model is actually refined for a particular application.

So far, we have looked at two of the diagrams in some detail when class diagrams and statecharts were used to illustrate static and behavioural modelling; these diagrams are shown again in this chapter for the sake of completeness and also to introduce the meta-model using diagrams that are already well known.

The order in which the diagrams are introduced in this chapter has no real rhyme nor reason. They are simply presented in, what is from the author's point of view, a logical order. Therefore, the various parts of this chapter may be read in any order.

This chapter also uses a worked example, which is a game of chess. There are several reasons for choosing this as the subject of the example:

- Most people have a good understanding of the game of chess, yet very few people can actually agree on what this understanding is.

- Chess is a game that can be realised in a number of different ways, such as: in software, in hardware, as an off-the-shelf board game, or by using helicopters and a field (apparently). Thus, it is interesting to see how a high-level model may be common to all, but, as time goes on and the model approaches reality, the models will start to differ.

- There are many aspects that can be modelled in the game of chess, from the board and pieces, to the rules of the game, and to strategies and skills in playing the game.

Now that the scene has been set, it is time to plunge straight in and look at the UML diagrams themselves.

5.2   Class diagrams

5.2.1   Overview

This section introduces the first of the nine UML diagrams: the class diagram. The class diagram was introduced in Chapter 3 in order to illustrate static modelling, so this section should really only serve as a recap. It is useful to examine class diagrams again, however, as they are the diagram that is used in order to realise the meta-model.

Class diagrams realise the static model of a system and show what 'things' exist in a system and what relationships exist between them. The 'things' in a system are represented by classes and their relationships are represented, unsurprisingly, by relationships.

Class diagrams are used to visualise conceptual, rather than actual, aspects of a system. Classes themselves represent abstractions of real-life entities. The actual real-life entities are represented by objects and, therefore, any object must have an associated class that defines what it looks like and what it does.

5.2.2 Diagram elements

Class diagrams are made up of two basic elements: classes and relationships. Both classes and relationships may have various types and have more detailed syntax that may be used to add more information about them. However, at the highest level of abstraction, there are just the two very simple elements that must exist in the diagram.

Classes describe the types of 'things' that exist in a system, whereas relationships describe how these classes are related together.

Figure 5.1 shows a high-level meta-model of class diagrams. Ironically, the meta-model for class diagrams is created using a class diagram, which makes it a class diagram of a class diagram. All future meta-models in this book will also be class diagrams.

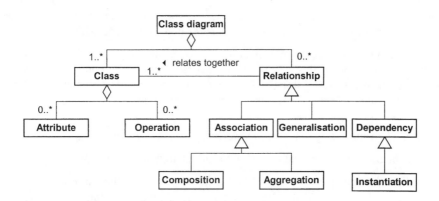

Figure 5.1 *Partial meta-model for class diagrams*

Remember that the UML is a language and, therefore, should be able to be read and understood just like any other language. Therefore, the model in Figure 5.1 may be read by reading class name, then any association that is joined to it, and then another class name.

Therefore, reading the diagram: a 'Class diagram' is made up of one or more 'Class' and zero or more 'Relationship'. It is possible for a class diagram to be made up of just a single class with no relationships at all; however, it is not possible to have a class diagram that is made up of just a relationship with no classes. Therefore, the multiplicity on 'Class' is one or more, whereas the multiplicity on 'Relationship' is zero or more.

Each 'Relationship' relates together one or more 'Class'. Notice that the word 'each' is used here, which means that for every single 'Relationship' there are one or more 'Class'. It is also interesting to note that the multiplicity on the 'Class' side of the association is one or more, as it is possible for a 'Relationship' to relate together one 'Class' – that is to say that a 'Class' may be related to itself.

Each 'Class' is made up of zero or more 'Attribute' and zero or more 'Operation'. This shows how classes may be further described using attributes and operations, or that they may be left with neither.

There are three main types of 'Relationship':

- 'Association', which defines a simple association between one or more "Class". There are also two specialisations of 'Association': a special type of 'Association' known as 'Aggregation' and 'Composition', which show 'is made up of' and 'is composed of' associations.

- 'Generalisation', which shows a 'has types' relationship that is used to show parent and child classes.

- 'Dependency', which has only one type (of many) shown, called 'Instantiation', which describes an 'is an example of' relationship.

That, in a nutshell, is the meta-model for class diagrams. Notice how the meta-model, even at this very high level, describes all the concepts that were introduced in Chapter 3. Imagine how this model may be expanded upon to show that, for example, each attribute has a type, default value, etc.

5.2.3   Example diagrams and modelling – class diagrams

Class diagrams may be used to model just about anything, including:

- Modelling physical systems. Many examples of modelling physical systems are used throughout this book, ranging from modelling cats and swords to modelling board games!

- Process modelling. Understanding processes, whether they be in standards procedures or someone's head, is very important in the world of systems engineering. The UML is amazingly powerful for modelling such processes and is the topic of Chapter 6.

- Teaching and training. The UML is a very good tool for teaching and training. The fact that you are reading this book and (hopefully) understanding the concept associated with class diagrams means that the UML is a good and effective training tool. Remember, more than anything, that the UML is a means to communicate effectively, which is what training and teaching are all about.

- Software. The UML is also very useful for modelling software, which is covered in enormous detail in just about every other UML book that exists and therefore will not be covered in much detail in this one!

Figure 5.2 shows that a 'Chess game' is made up of two 'Player' and that each player has attributes – 'result' and 'status' – and is able to perform an action called 'move'.

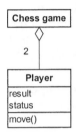

Figure 5.2   *High-level class diagram for the chess game example*

This was the example chosen to illustrate static and behavioural modelling, which was covered in previous chapters, so let us now consider another aspect of chess that has been completely omitted from the model shown above, but without which the game cannot be played: the board. The next section, therefore, considers the board itself and how this may be modelled. It has been mentioned before that there are many correct solutions to a single problem, which will be illustrated over the next few diagrams.

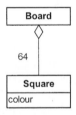

Figure 5.3   *Model of a chessboard*

Figure 5.3 shows a very simple representation of a chessboard. It can be seen that 'Board' is made up of 64 'Square'. Each 'Square' has a single attribute called 'colour', the value of which will dictate which colour the square will be on the board. This is a very simple model and is correct. It is may not be the best representation and does not contain much detail, but it is nonetheless correct.

It is interesting to see how more detail could be added, such as, is it possible to state that there are two, and only two, possible values for the attribute 'colour', being black or white. Also, is the aggregation symbol correct, or would a composition symbol be more accurate? This will depend on the nature of the board and it is necessary to ask the question of whether each square can actually exits by itself or whether it is, inherently, part of the board. In a typical shop-bought chess set, the symbol is more appropriate if it is a composition symbol, rather than aggregation. However, what about if each square were actually a field and the pieces helicopters? This may seem slightly ridiculous, but it serves to illustrate the point that almost every model can be represented slightly differently and these are the type of question that any modeller must begin to ask themselves.

The previous points were slight modifications of the existing model, but it is also possible to have a very different-looking model that still represents the same information, as shown in Figure 5.4.

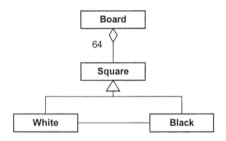

Figure 5.4   *Alternative chessboard model*

Figure 5.4 presents an alternative interpretation of the chessboard. This time, the 'Board' is still made up of 64 'Square' (aggregation or composition?) but the class 'Square' has no attribute to represent colour. By changing the model, it is possible to show that the concept of colour still exists, but this time by defining two specialisations of 'Square': 'Black' and 'White'. More information is added by showing that there is an association between 'Black' and 'White', but note the multiplicity on the association. As there is no multiplicity defined, one assumes that it is a one-to-one relationship. This implies that, not only are there 64 'Square', but there is a one-to-one relationship between the two types of 'Square': 'Black' and 'White'. Or to put it another way, there must be 32 'Black' and 32 'White' in order for the association to be true.

The model was changed by adding more classes, but more information was added by doing this. The model has become more complex as there are more classes and associations on it.

The final example of the chessboard is shown in Figure 5.5, illustrating a more radical change of model.

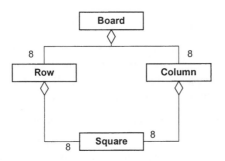

Figure 5.5 *Final example of the chessboard model*

Figure 5.5 shows the chessboard again, but this time the whole pattern has changed, as the basic structure of the aggregation has changed. This time, 'Board' is made up of eight 'Column' and eight 'Row'. Each 'Column' and each 'Row' is made up of eight 'Square'. There is no concept of colour here, but this could be addressed in both approaches shown previously: attributes or subclasses. In addition, there is the same issue of aggregation versus composition here.

The question that most people ask at this point is, 'which is the best model?' The question that should be asked, however, is 'which is the most appropriate model for the problem at hand?' Depending on the criteria for defining what 'best' means, the most appropriate model will change. Imagine if the problem called for the simplest (least complex) of models, the most appropriate would be Figure 5.3. If the criteria were more concerned with the layout of the squares on the board, then perhaps Figure 5.4 or Figure 5.5 would be more appropriate.

All the models shown previously in this section have described the same piece of information and this was very simple information. Imagine trying to represent a slightly more complex problem, such as the chess pieces, where there are still two colours but also six different types of piece, some of which exist singularly, some in pairs and some in sets of eight. This is still a relatively simple problem. Now imagine scaling the problem right up to represent an automated chess machine that can also teach chess, tell the time and make coffee – imagine how many different possible models would exist there!

The main point that has to be made here comes back, once more, to communication. It does not matter how many or few models are generated, because if they cannot be communicated effectively to other people, then how is it possible to make an informed and intelligent decision as to which one is the most appropriate for the problem at hand? The UML offers this method of communication that can iron out many of the ambiguities in the form of a group decision, rather than it being made by a single person.

5.2.4   Using class diagrams

Class diagrams are used to model just about anything and form the backbone of any UML model. Class diagrams are perhaps the richest in terms of the amount of syntax available and, because of this, the meta-model for class diagrams is the most incomplete. Other features of class diagrams include, but are not limited to:

- Extra types of classes, such as association classes and interface classes.

- Extra types of relationships, such as other types of dependencies and extra detail that can be added to relationships, such as role names and qualifiers.

The main aim of the class diagram, as with all UML models, is clarity and simplicity. Class diagrams should be able to be read as any sentence and they should make sense. Diagrams that are difficult to read probably represent complex information and, as such, should be simplified or, if this is not possible, then lessons should be learned from them.

5.3   Object diagrams

5.3.1   Overview

This section introduces the second of the nine UML diagrams: the object diagram. Object diagrams realise part of the static model of a system and are heavily related to class diagrams. In fact, many people confuse the terms 'object diagram' and 'class diagram' and use them interchangeably, but they are in fact two separate diagrams.

The definition of an object is that it is an instance of a class; therefore, it seems logical that object diagrams will represent instances of class diagrams, which is exactly what they do. Objects represent real-life examples of the abstract classes found in class diagrams and relate very closely to real life. Object diagrams also have strong relationships with component and deployment diagrams (both static) as well as both types of interaction diagram (collaboration and sequence).

5.3.2   Diagram elements

Object diagrams are made up of two main elements: objects and links. Objects are defined as an instance, or many instances, of a class. In the same way, links are instances of associations. The model shown in Figure 5.6 shows the partial meta-model for object diagrams – bearing in mind that there is a lot more syntax available when using object diagrams and a lot more detail may be added to each class shown on this model.

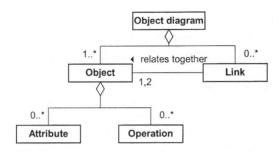

Figure 5.6   *Partial meta-model for object diagrams*

Object diagrams are very heavily dependent on class diagrams, as each object requires an associated class from which its appearance and behaviour can be taken. This is illustrated in Figure 5.7, which shows a meta-model of how classes and objects are related.

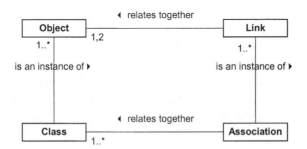

Figure 5.7   *Meta-model showing relationship between classes and objects*

Reading the diagram, it can be seen that a 'Link' relates together two 'Object'. One or more 'Object' is an instance of a 'Class' and one or more 'Link' is an instance of an 'Association'.

Notice how the top and bottom half of Figure 5.7 have both been seen before in Figure 5.1 and Figure 5.6. This is an example of seeing patterns emerge from different models that can give a whole new view of a particular piece of information when shown together and with the addition of a few extra relationships [7].

5.3.3   Example diagrams and modelling – object diagrams

Object diagrams are perhaps the least used of all the nine UML diagrams and, indeed, many of the tools on the market do not support the creation of object diagrams. However, like all aspects of life, it is difficult to see how some things can be used effectively until a particularly awkward situation arises that cannot be modelled using any other diagram. This section aims to point out some real-life examples of how object diagrams can be very useful, if only to illustrate how a very abstract class diagram relates to reality. In other words, object diagrams are very good at making that all-important connection to reality that all good models should have. This will be illustrated by looking at two real-life examples of how useful an object diagram can be, before relating object diagrams back to the chess example.

Consider the example of a taught course and how different types of person may exist who are involved with the course. Each of these people may have attributes that make them unique and that may be useful to administration staff who run the course. This may be modelled by the class diagram shown in Figure 5.8.

It can be seen from Figure 5.8 that there is a class called 'Person' that has five attributes to describe that person: 'name', 'company', 'phone', 'email' and 'fax'. There are two types of 'Person' – 'Tutor' and 'Attendee' – each of whom inherit the attributes of their parent class. 'Tutor' is specialised by having the operation 'impart', while 'Attendee' has operations 'listen' and 'learn'.

It is now possible to show an example of some people who may have their records stored on a very simple database. Although the class diagram shown here is a far cry from a database design, it clearly illustrates the point that sometimes it is useful to see an example of how such static information may be stored.

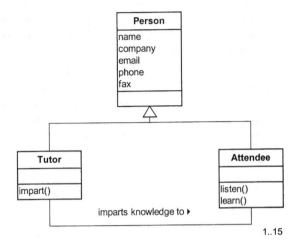

Figure 5.8 *Class diagram of a taught course*

: Tutor	imparts knoweldge to ▶	: Attendee
name = Jon Holt company = Brass Bullet Ltd email = jon@brass-bullet.co.uk phone = 01792 417227 fax = 0870 055 8729		name = Joe Blob company = Rubber Monkey Ltd email = joe@rubber-monkey.co.uk phone = 0207 999999 fax = 0207 999999

Figure 5.9 *Example of an object diagram*

Figure 5.9 shows examples of two of the records that may exist: one for 'Tutor' and one for 'Attendee'. Notice how the objects here are anonymous instances as they do not have unique object names identified for them. It is assumed that each will actually have a unique identifier, but for whatever reason they are not shown here. The two objects are clearly instances of their associated classes, but the original association from Figure 5.8 has now been instantiated as a link that shows the relationship between the two objects. Notice how the multiplicity of the link is different from that of the association as the association is abstract, showing a 1 to 1...15 multiplicity. The link, however, shows a 1 to 1 multiplicity, which is still correct as the original multiplicity means between one and 15. If the diagram were completed, there would be between one and 15 instances of 'Attendee' each with a 1 to 1 relationship with the instance of 'Tutor'.

Object diagrams, it has been stated, represent real-life examples of their associated classes and thus have their attribute values filled in. These may never change once set, such as the malignly monickered 'Joe Blob', or they may change over time, as in the chess example later in this section.

Another example of the use of object diagrams is when the class diagram is a little too abstract to get a clear handle on what is happening. This is often the case when massive amounts of data and information are abstracted into a very simple hierarchy, as often happens in databases. The following example is based on a system that must describe how different components stored in a database are configured to make a whole assembly.

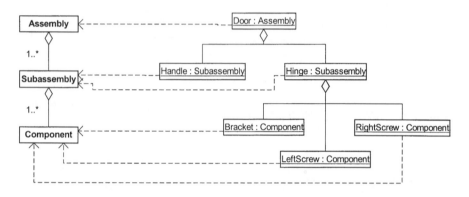

Figure 5.10 *Combined class and object diagram for a parts database*

Figure 5.10 shows a combined class and object diagram, which is one that should usually be avoided. However, in this case, it shows the link relationship between a very abstract class diagram and a real-life object diagram quite nicely.

The classes on the left of the model show a simple hierarchy that represents the way that parts in a database are classified before they are configured to make up a complete system. The highest level in the hierarchy is 'Assembly', which represents a completed unit that may be sent out of a warehouse to be used in the manufacture of vehicles, for example. It can be seen that each 'Assembly' is made up of one or more 'Subassembly', each of which is made up of one or more 'Component'. This class hierarchy is very easy to understand in terms of the class diagram syntax itself, but is very difficult to relate to a real-life situation. It does not matter how accurate the class hierarchy is if the person paying the bill cannot make head nor tail of it; therefore, the obvious solution would be to show an example of the real-life situation in which it would appear. This is where the object diagram comes in.

Consider the right-hand side of the diagram now, which makes up the object diagram. It can be seen that an instance of 'Assembly' called 'Door' is made up of two instances of 'Subassembly': 'Handle' and 'Hinge'. 'Hinge' is further decomposed into three instances of 'Component': 'LeftScrew', 'RightScrew' and 'Bracket'.

Although very simple, it was necessary to show an example hierarchy to the customer otherwise there is no way that they could agree that the class diagram was correct as it was too abstract for the customer to understand. To explain this another way, the connection to reality was too far removed and thus the customer had problems understanding it. Obviously it is simply not feasible for the object diagram to show every instance of the class as it would be far too large to contemplate.

The following model continues the chess example that was introduced previously, but this time the object diagram is used.

Figure 5.11 *Object diagram for the chess game*

Figure 5.11 shows one possible object model for the chess game. Notice how each object holds true to the original classes that were introduced in Figure 5.2. This time, however, as we are considering actual objects, the attribute values are filled in. This may change throughout the lifetime of the object: as the object changes state so might the value of the attributes.

In addition, note that a link has been included that is an instance of the association from the class diagram. Also, note that the multiplicity here is one to one, which is the same as in the original class diagram.

There is a strong link between object diagrams and statecharts, as statecharts describe the behaviour over the lifetime of a class, which is the same as saying an object. Therefore, any state in an object's statechart should be able to be represented by an object diagram that reflects the object's attribute values while it is in that state.

5.3.4 Using object diagrams

There are a few hints to bear in mind when using object diagrams:

- Each object must be an instance of a class and there may be more than one object (instance) for every class. If an object needs to be represented on a diagram and a class does not already exist, it is necessary to revisit the class diagrams in the model and create an appropriate class.

- Objects exist in the real world as specific examples (instances) of things (classes). As a consequence of this, each object is unique and has a unique identifier. In many cases, this identifier need not be shown on the model but it should be remembered that it does exist.

- Each link is an instance of an association and one association may have more than one link as its instances. Again, this can provide a good consistency check with the associated class diagram.

- Attributes in an object have their values filled in, to reflect their 'real' nature. An object with its attributes' values filled in should correspond to at least one state in that object's statechart.

- Objects can also be seen on interaction diagrams (both types) and deployment diagrams. An interaction diagram is basically a behavioural view of the object diagram, which will show messages on links and ordering information.

Object diagrams are one of the least used of all the UML diagrams and they are not supported by many CASE (computer aided/assisted software engineering) tools. Despite this, however, they do still have their uses, as described in this chapter.

5.4   Statechart diagrams

5.4.1   Overview

This section introduces the third of the nine diagrams: the statechart. Statecharts have been discussed in some detail in Chapter 4 and thus some of this section will serve as a recap. The focus here, however, will be the actual statechart diagram, whereas the emphasis previously has been on general behavioural modelling.

Statecharts are a type of behavioural model and are sometimes referred to as 'timing models', which may, depending on your background, be a misnomer. The term 'timing' here refers purely to logical time, rather than real time. By 'logical time' here it means the order in which things occur and the logical conditions under which they occur. There is no inherent mechanism for representing real time in statecharts.

This lack of real-time capability is often heralded as one of the shortfalls of the UML but, technically speaking, this is not actually true. The UML has never claimed to be a real-time modelling technique and thus these criticisms fall outside the scope of the UML. The question of real time, however, should not be as neatly side-stepped as this and, indeed, there are several extension mechanisms for the UML that claim full real-time capabilities, such as determinism and full temporal modelling. See Douglass' work for more details of real-time modelling using the UML [4].

Statecharts model the behaviour of an object or, as it is sometimes written, they model the behaviour during the lifetime of a class, which is the same thing. It has already been stated that statecharts model the order in which things occur and the conditions under which they occur. The 'things' in this context are, broadly speaking, activities and actions. Activities should be derived from class diagrams as they are equivalent to operations on classes. Indeed, one of the basic consistency checks that can be carried out is to ensure that any operations on a class are represented somewhere on its associated statechart as activities or actions.

5.4.2 Diagram elements

Statecharts are made up of two basic elements: states and transitions. These states and transitions describe the behaviour of an instance of a class over logical time. States show what is happening at any particular instance in time when an object is active. States may show when an activity is being carried out or when the attributes of an object are equal to a particular set of values. They may even show that nothing is happening at all – that is to say that the object is waiting for something to happen.

Figure 5.12 shows the partial meta-model for statecharts. Statecharts have a very rich syntax and thus the meta-model shown here omits some detail – for example, there are different types of action that are not shown.

From the model, it can be seen that a 'Statechart diagram' is made up of one or more 'State' and one or more 'Transition'. A 'Transition' shows how to change between one or two 'State'. Remember that it is possible for a transition to exit a state and then enter the same state, which makes the multiplicity one or two rather than two, as would seem more logical.

There are three types of 'State': 'Normal', 'Start' and 'End'. Each 'Transition' may have zero or more 'Condition', which explains how a transition may be crossed. A condition is a Boolean condition that will usually relate to the value of an attribute.

A 'Transition' may also have zero or more 'Action'. An action is defined as an activity that takes no time, and one that, depending on one's viewpoint, could be perceived as impossible. In software terms, this may be a very quick activity, such as opening a window.

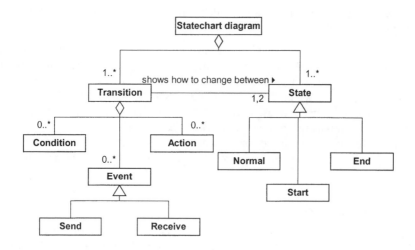

Figure 5.12 *Partial meta-model for statecharts*

Finally, a 'Transition' may have zero or more 'Event'. An event is, basically, the passing of a message, usually from one object to another, or to put it another way, from one statechart to another.

There are two main types of 'Event': 'Send' and 'Receive'. A 'Send' event represents the origin of a message being sent from one statechart to another. It is generally assumed that a send event is broadcast to all objects in the system and thus each of the other objects has the potential to receive and react upon receiving the message. Obviously, for each send event there must be at least one corresponding receive event in another statechart. This is one of the basic consistency checks that may be applied to different statecharts to ensure that they are consistent. This is analogous with component and integration testing, as it is possible (and, in some cases, difficult to avoid) to completely verify the functional operation of an object, yet when it comes to integrating the objects into a system, they do not interact correctly together. Indeed, this was the case in the chess example used in Chapter 4.

Perhaps the simplest and most common source of errors can be avoided by applying a simple consistency check between the statechart and its associated class. This relationship can be seen in Figure 5.13.

Figure 5.13 shows the relationship between a statechart and its associated class. From the model, one or more 'Statechart diagram' shows the behaviour over the lifetime of a 'Class'.

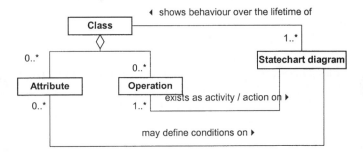

Figure 5.13 *Meta-model relationship between classes and statecharts*

Each 'Operation' in a 'Class' must exist as an activity or action on a 'Statechart diagram'. Each 'Attribute' in a 'Class' may define a condition on a 'Statechart diagram'. Any conditions that are defined should be related back to an attribute of the class, which adds another consistency check to the model.

A simple rule of thumb that will go some way to ensuring that all classes in a system are fully defined, is that to say any class that exhibits behaviour (that has operations) must have an associated statechart.

5.4.3   Examples and modelling—statecharts

This section looks at some practical examples of modelling and revisits the chess game statechart that was used previously. This time, however, the emphasis is on the actual mechanics of using statecharts, rather than on the behavioural concepts that were covered previously.

Figure 5.14 shows the statechart for the class 'Player' from the chess example.

Three types of state are shown here:

- A 'Start' state, realised by a filled-in circle that indicates the creation of an instance of the class.

- An 'End' state, realised by two concentric circles that indicates the destruction of the instance of the class.

- Normal states, indicated by rounded boxes, show what is happening in the object at any particular point in time. It is important to ensure that the rounded boxes actually have vertical straight sides so that it is just the corners that are rounded. The reason for this is that activity diagrams (which are actually a special type of statechart) have two types of state – action states and activity states – and the only visual difference between these two and a normal state is that action and activity states have no straight verticals whereas normal states do. Mixing these types of state can lead to a great deal of confusion, even in statecharts!

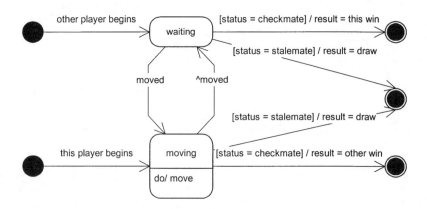

Figure 5.14 *Statechart for the 'Player' class*

Normal states may or may not have activities or actions associated with them that should correspond directly to operations from the class diagram. Each operation from a class must exist as an activity on its associated statechart and, likewise, any activities that exist on the statechart must be present as operations on the governing class. Examples of both are shown here, where the 'moving' state has the 'move' activity, while the 'waiting' state has no activity.

The only way to go from one state to another is to cross a transition. Transitions are realised by directed lines that may have events and conditions on them. Transitions are uni-directional and if a return path of control is required, so too is another transition.

Reading the diagram, the object can begin its life in one of two ways: by the event 'this player begins' or 'other player begins'. If 'other player begins', the first state is 'waiting'. When the 'moved' event occurs, the object enters the moving state when 'move' is executed.

If 'this player begins', the first state is 'moving' where 'move' is executed. When the send event '^moved' occurs, the object enters the 'waiting' state. The '^' before the event 'moved' indicates that this is a send event, rather than a receive event. Another way to differentiate between send and receive events is to, rather cunningly, write the word 'send' before an event name to indicate that it is indeed a send event.

There are three possible ends to the game, represented by the three end states:

- If 'status = checkmate' while the object is 'waiting', the attribute 'result' is set to 'this win'.

- If 'status = checkmate' while the object is 'moving', the attribute 'result' is set to 'other win'.

- If 'status =stalemate' from either 'moving' or 'waiting', 'result' is set to 'draw'.

Conditions are used to describe the difference between each transition as a state cannot have two identical conditions associated with it. The condition is a Boolean expression and, in this case, the left-hand side of the condition relates directly to an attribute. This provides rather a nice consistency check back to the statechart's governing class.

Note the use of actions here to set the attribute value to one of its pre-determined values. Actions may be related to operations in a class diagram or may even refer to a simple assignment, or very low-level aspect of a design or specification. Remember again the link back to object diagrams and imagine an object with its attribute values filled in, which would correspond to this state.

The statecharts used so far have had an emphasis on using states where something happens over a period of time that is represented as an activity. There is another approach to creating statecharts that is more event-driven and that uses far more actions than activities. Indeed, in many such statecharts there are no activities at all, only actions. An example of an activity-driven statechart can be seen in Figure 5.15.

Figure 5.15 shows the statechart that represents the lifetime of an object. It can be seen here that the state name does not have the 'ing' ending as, this time, it is representing the state when particular conditions are met, rather than when something is actually happening. The things that happen here are shown as actions on the transitions between states.

There is a fine line between whether something that happens should be an action or an activity, but the following guidelines should give some pointers as to which are which:

- Actions are assumed to be non-interruptible – that is, that they are atomic in nature and once they have been fired, they must complete their execution. Activities, on the other hand, may be interrupted once they are running.

- As a consequence of the first point, actions are assumed to take zero time! This, however, refers to logical time rather than real time, which reflects the atomic nature of the action. Activities, on the other hand, take time to execute.

- Activities may be described using activity diagrams, whereas an action, generally speaking, may not.

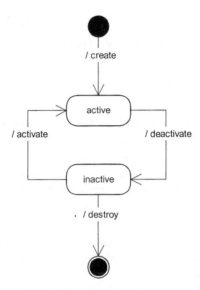

Figure 5.15 *Event-based statechart representing the lifetime of an object*

It is important to differentiate between these activities and actions as they can have a large impact on the way in which the system models will evolve.

5.4.4   Using statecharts

There are a few rules of thumb to apply when creating statecharts:

- All classes that exhibit behaviour (have operations) must have an associated statechart, otherwise the system is not fully defined. This reflects the fact that there must always be the two views of the system and forms a fundamental consistency check for any model.

- All operations in a particular class must appear on its associated statechart. States may be empty and have no activities, which may represent, for example, an idle state where the system is waiting for an event to occur. Messages are sent to and received from other objects, or statecharts.

One question that is often asked concerns which of the two approaches to statecharts is the better: the activity approach or the event-driven approach? There is not much difference between them as it really is a matter of personal preference. However, having said this, the consistency checks for an activity-driven statechart are far stronger that the activity-driven statechart, as the activities are related directly back to the governing class. Many books will only use examples where activity-driven statecharts are considered, rather than both types. It should also be borne in mind that it is easier to transform an activity into an action, that it is to transform an action into an activity. Therefore, when creating statecharts, it is suggested that activities are used on the first iteration of the model.

5.5   Interaction diagrams

5.5.1   Introduction

This section takes a bit of a departure from the other sections in this chapter, as two diagrams are being discussed rather than one. This is because an interaction diagram can be illustrated using two different types of diagram: collaboration and sequence. Both diagrams show the same information, but from two slightly different viewpoints. This relationship can be seen in the simple meta-model shown in Figure 5.16.

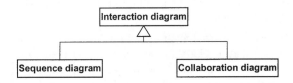

Figure 5.16 *Meta-model showing types of interaction diagram*

The two types of interaction diagram do differ, but the discussion concerning these differences will be saved until the end of this section.

5.5.2   Overview – collaboration diagrams

This section introduces and discusses collaboration diagrams. Collaboration diagrams realise a behavioural view of a system. A collaboration diagram models interactions between instances of classes, or objects.

The main aim of a collaboration diagram is to show a particular example of operation of a system, in the same way as movie-makers may draw up a storyboard. A storyboard shows the sequence of events in a film before it is made. Such storyboards in the UML are known as 'scenarios'. Scenarios highlight pertinent objects in a particular situation and ignore all others.

Effectively, the collaboration diagram shows message flows between objects, which are displayed in terms of their physical organisation.

5.5.3   *Diagram elements – collaboration diagrams*

Collaboration diagrams are made up of three basic elements: objects, links and messages.

Objects have already been introduced and were used in, unsurprisingly, object diagrams. However, whereas the object diagram realised a static view of the system, showing the values of attributes at a particular time instance, collaboration diagrams realise a behavioural view of the system, concentrating on the messages that are passed between objects. In the same way, links have also been introduced previously as being instances of associations. Again, however, the emphasis previously has been on static links, whereas the emphasis now is to show the messages that are passed along the links. This brings us rather neatly on to messages, which have been introduced previously as events that were passed between statecharts.

The more modelling that is carried out, the stronger the links between the various types of diagram. It is this strong link between diagrams that enables a good, correct, consistent design to be produced, and one in which the designer can have a great deal of confidence.

Figure 5.17 shows the partial meta-model for collaboration diagrams. It can be seen from the model that a 'Collaboration diagram' is made up of two or more 'Object', one or more 'Link' and one or more 'Message'. The multiplicity between 'Collaboration diagram' and 'Object' is '2..*', because although it is possible to have a link from one object back to itself, there would be little or no value showing a collaboration diagram with a single object. The word 'collaboration' implies that two or more entities are working together, or collaborating.

A 'Link' shows the relationship between one or two 'Object' and a 'Message' describes a 'Link'. Again, it is possible to show a link going from one object to itself, and thus the multiplicity is '1,2', rather than simply '2'.

Figure 5.18 shows the relationships between collaboration diagrams and how their elements relate to other UML elements.

Collaboration diagrams are closely related to object diagrams, which, in turn, are closely related to class diagrams. Therefore, from the model, an 'Association' relates together one or more 'Class'. One or more 'Object' is an instance of a 'Class' and one or more 'Link' is an instance of an 'Association'. Therefore, a 'Link' relates together one or two 'Object'.

Another relationship that is shown here is that one or more 'Collaboration diagram' describes an instance of a 'Use case'. A scenario, in UML terms, is defined as an instance of a use case, hence the relationship between 'Use case' and 'Collaboration diagram'. This is discussed in more detail in Chapter 7.

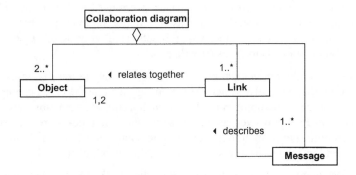

Figure 5.17 *Partial meta-model for collaboration diagrams*

Notice how familiar patterns are beginning to emerge from this model. The lower part of the model was seen in both the class meta-model (Figure 5.1) and the object diagram meta-model (Figure 5.6) and then once again when showing the relationships between class and object diagrams. The upper half of the model is taken directly from the object diagram meta-model.

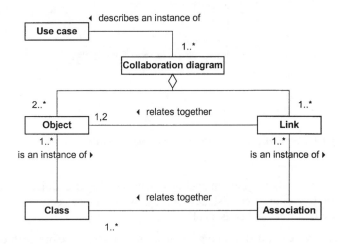

Figure 5.18 *Meta-model showing relationship between collaboration diagrams and other UML elements*

5.5.4 Examples and modelling – collaboration diagrams

Collaboration diagrams are very useful for demonstrating the consistency of a set of models. In particular, they are crucial for ensuring that the behaviour of the system, at a high level, is correct. Statecharts were used to define the behaviour within objects and it was stated that it was very important to ensure that the events that are passed between statecharts (objects) are consistent. One use for collaboration diagrams is to do exactly this.

Let us revisit the chess example and see how collaboration diagrams may be used to model that particular system. It may be argued that the chess example is so simple that there is little point in modelling such a simple interaction between two such simple objects. However, as will be seen when the models are created, there has been an omission from all our previous chess game models, which only comes to light when the behaviour of a system is modelled at a higher level.

The class diagram for the game of chess indicated that each 'Chess game' was made up of two 'Player'. As the multiplicity is explicit, two instances of 'Player' are created and the behaviour of each instance is dictated by the statechart for 'Player'. The game begins with the creation of two objects – let us call them 'Player 1' and 'Player 2'. There is clearly going to be an interaction between the two objects that will be shown as a link. The message that is passed along this link will be the same as the events passed between the statecharts. From this information, the model in Figure 5.19 can be drawn up.

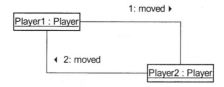

Figure 5.19 *Collaboration for the chess game example*

Figure 5.19 shows the interactions between the two objects. Let us now perform a few simple consistency checks on the model so far to ensure that everything is correct and consistent.

Objects were defined as being instances of classes, therefore each object must have a class from which it takes its appearance and behaviour. A quick look at the class diagram in Figure 5.2 proves that this aspect of the model is consistent.

The message that is passed along the link has already been defined in the statechart for the two objects, so this may be verified as being consistent and correct.

Links were defined as being instances of association, therefore each link must have an association upon which it is based. Another quick look at Figure 5.2 shows that no such association exists. This means that the model is incorrect, as there is an association missing from the original class diagram. The class diagram, therefore, must be modified in order to bring it in line with the rest of the model. This modified class diagram is shown in Figure 5.20.

Figure 5.20 shows the class diagram from Figure 5.2 but with an additional association. The association is called 'moves against' and provides the template required for all its instances, one of which is represented by the link that can be seen in the collaboration diagram. Note that the association is bi-directional as no direction is explicitly stated, which will translate into two uni-directional links when realised in the collaboration diagram. Note also, that as no multiplicity is indicated, it is assumed to be a '1 to 1' relationship.

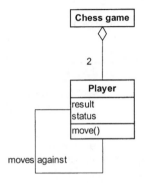

Figure 5.20 *Updated class diagram for the chess game*

This has highlighted an essential feature of modelling that was first touched upon in the previous chapter, that of levels of abstraction of models. Both statecharts and collaboration diagrams are types of behavioural model, but they represent different levels of abstraction of the model. It was only when the behavioural model was looked at from both levels that an error was thrown up. Collaboration diagrams represent a high level of abstraction of any model, whereas a statechart shows a lower level of behaviour. It is vital to ensure that consistency checks are performed on, and between, models of the system.

This has also highlighted another crucial feature of the UML, which is that no matter how simple the model, it is almost impossible to get it right on the first attempt. Achieving a good consistent model requires a great deal of iteration between various types of diagram and the more modelling that takes place, the more consistent the model will become and the closer the connection to reality will be.

Another way to use collaboration diagrams, which is the use suggested by almost all UML literature, is to model scenarios. In reality, this is exactly what has just been discussed, but with a potentially narrower scope.

As an example of modelling scenarios, one example that will be looked at in detail in due course is 'process modelling'. The processes themselves are represented statically by classes and thus behaviourally by statecharts. Therefore, the way in which each process is described behaviourally has already been defined, but the way in which they interact with each other may be configured differently. An example of this is the partial life cycle model that is shown in Figure 5.21, which is based on the configuration of well-defined processes and which may be modelled very well using a collaboration diagram. Process modelling is discussed in more detail in Chapter 6.

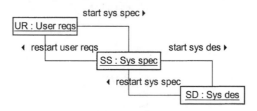

Figure 5.21 *Collaboration diagram showing a project lifecycle model*

5.5.5   Using collaboration diagrams

There are several rules of thumb that should be borne in mind when creating collaboration diagrams:

- Objects in a collaboration diagram will always be instances of classes. This is true whether the classes have been defined or not. If the classes have been defined, all is well and good. If, however, classes have not been defined, it is a good approach to identifying classes based on example scenarios (perhaps created from particular use cases).

- In the same way, links are always instances of associations. This can prove to be a good consistency check back to the class diagram.

- Messages may be added to links that may be derived from statecharts – if they already exist. If not, each message may show which send and receive events may be required for statecharts.

- One final point that should be borne in mind is that collaboration diagrams model behaviour according to organisation.

Collaboration diagrams are typically used to model scenarios; however, there is another use that is rarely mentioned in the literature, which is that of a consistency check for objects and their associated statechart. This point was raised in Chapter 4 when the statechart for the class 'Player' was able to be verified as working at one level, but it took a higher level view to reveal that the messages defined on the statechart were incorrect. This also relates to Chapter 2 where it was stated that models should be looked at from more than one level of abstraction to ensure that the overall model is correct. Collaboration diagrams (and sequence diagrams) are very powerful when used as a consistency check between various interacting objects that already have their internal behaviour defined with statecharts.

It should be stressed again that logical timing relates purely to the order in which things happen and does not imply any real-time information.

5.5.6   Overview – sequence diagrams

This section introduces and discusses sequence diagrams, which realise a behavioural view of the system. Sequence diagrams are one of the two types of interaction diagram and, as such, behave very much like collaboration diagrams. Indeed, the information shown by both types of interaction diagram is actually the same, but from a slightly different point of view. As such, sequence diagrams show objects, links between objects and the messages that are passed between objects.

The main difference between a collaboration diagram and a sequence diagram is that collaboration diagrams show the high-level behaviour of a system from an organisational point of view, whereas sequence diagrams show the high-level behaviour of a system from the logical timing point of view.

5.5.7   Diagram elements – sequence diagrams

Sequence diagrams are composed of three basic elements – objects, links and messages – which are exactly the same as for the collaboration diagram. However, the objects shown in a sequence diagram have a lifeline associated with them, which shows a logical time line, i.e. it only shows the order or events (logical timing dictates the order). This time line is present whenever the object is active or alive, and is represented graphically as a vertical line with logical time travelling down the line. Another difference between the diagrams is the focus of control that defines, basically, when the object is active and when it is idle. These foci of control should be consistent with statecharts in terms of active and idle states.

The objects for the sequence diagram are shown going horizontally across the page. They should, theoretically, be shown staggered down the diagram dependent on when they are created. In reality, however, it is often the case that all objects are shown at the same level across the top of the diagram.

The partial meta-model for sequence diagrams in shown in Figure 5.22, which is very similar to the meta-model for the collaboration diagram, except for the 'Lifeline' and 'Focus of control' that have been added beneath 'Object'.

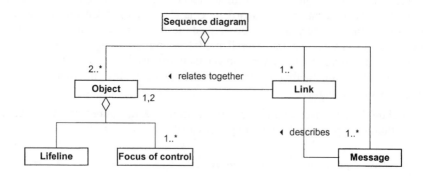

Figure 5.22   *Partial meta-model for sequence diagrams*

Figure 5.22 shows that a 'Sequence diagram' is made up of two or more 'Object', one or more 'Link' and one or more 'Message'. So far, this is the same as the meta-model for the collaboration diagram, which was seen in Figure 5.17. An 'Object' is made up of a single 'Lifeline' and one or more 'Focus of control'.

A 'Link' shows the relationships between one or two 'Object'. Each 'Message' describes a single 'Link'.

5.5.8   Examples and modelling – sequence diagrams

In order to emphasise the difference between collaboration diagrams and sequence diagrams, each of the examples that were used for collaboration diagrams will be used again, so that the two types of interaction diagram can be compared directly. The first example will, once again, be the chess example.

Figure 5.23 shows the sequence diagram for the chess game, which shows the same information as the collaboration diagram in Figure 5.19. The objects in the system are realised in the usual way: instance name, colon, class name, all underlined. Two objects are shown here: 'Player 1' and 'Player 2'. These will both be created at the same time – when the game of chess begins – and, as such, both objects appear at the same height on the diagram.

The lifeline for each object is shown by the dashed vertical line under each object. The focus of control for each lifeline is shown by a rectangle that covers part of the lifeline. Note how each object's focus of control alternates between being present and being absent.

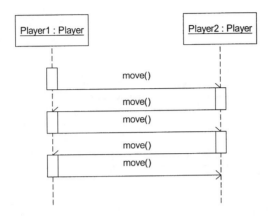

Figure 5.23 *Sequence diagram for the chess example*

In terms of how the game is played, it can be seen that 'Player 1' begins the game and, hence, the first focus of control is with 'Player 1'. All the time that 'Player 1' has an active lifeline (shown visually by the focus of control), 'Player 2' has an inactive lifeline (no focus of control present). The focus of control may now be related back to the original statechart to see if the models are consistent – remember that both models show behaviour but at different levels of abstraction. When 'Player 1' has an active lifeline, this relates to the 'Player 1' statechart being in the 'moving' state. Likewise, when 'Player 1' has an inactive lifeline, this relates to when the 'Player 1' statechart is in the 'waiting' state.

The focus of control for the game of chess passes from 'Player 1' to 'Player 2' once 'Player 1' has moved. This is indicated by the message 'moved' that is sent via the first link to 'Player 2'. 'Player 2' behaves in exactly the same way as 'Player 1' as they have the same statechart, hence the focus of control passes between 'Player 1' and 'Player 2' as the game progresses, until one player wins and the game is over. The end of the game and the end of both object's lives occurs when one player wins or a draw occurs. The end of the object's life is indicated by a large 'X' on the object's lifeline, which terminates the object. This is directly comparable with the end state from the original statechart.

The second example is that of the system life cycle model that was discussed in the collaboration diagram.

Figure 5.24 shows the sequence diagram for the life cycle model that was initially shown in Figure 5.21. This model shows the system life cycle again, but this time the emphasis is on the logical timing of the interactions.

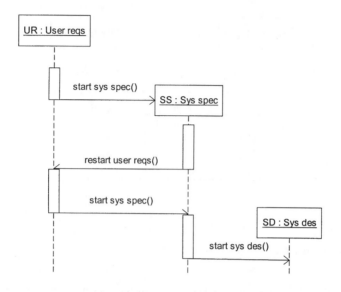

Figure 5.24 *Sequence diagram for a life cycle model*

From the model, the first thing that happens is that an instance of 'User requirements', called 'UR', is created, which has its own focus of control. When the focus of control shifts to the next object, the message will actually cause the creation of the second object, 'System spec', which is shown across from 'User requirements' but lower down the diagram, which indicates that it is created some time after the first object was created.

The next thing that happens is that the focus of control is passed back to the first object, 'UR', which indicates that, for some reason, the phase had to be re-visited. In reality, this would relate to an iteration of the system life cycle. When the focus of control passes back to 'SS', the lifeline becomes active once more. The focus then shifts to the third object, 'SD', which is created for the first time, and so forth.

5.5.9   Using sequence diagrams

The rules of thumb for the sequence diagrams are very similar to those for the collaboration diagram.

Both types of interaction diagram are semantically equivalent, in that they are based on the same information in the system. Indeed, it is a relatively simple matter to convert between them and this task may be performed automatically by some tools.

The main difference is that the emphasis is on logical timing in the sequence diagram. Collaboration diagrams give a better indication of how two objects are linked together but have weak logical ordering capabilities, as they are based on a simple numbering of events. Sequence diagrams, on the other hand, are weak at showing how objects are linked but stronger in showing the order in which things happen, including return paths of control.

An analogy that is often quoted for the two types of diagram is to think of Gantt charts and Pert charts in project management. A collaboration diagram is analogous to a Pert chart and a sequence diagram is analogous to a Gantt chart. Indeed, one use for sequence diagrams is to use them to plan Gantt charts as they show behaviour over logical time, which is a good basis for a Gantt chart.

5.5.10 *Sequence diagrams vs. collaboration diagrams*

It has already been stated that collaboration diagrams emphasise the organisational aspects of the system, while sequence diagrams emphasise the logical timing of the system. In practical terms, however, it can be seen that there are two distinct uses for the two types of diagram:

- Collaboration diagrams show how the objects may relate together and the objects may be grouped in any way in the model. However, there is very little logical timing information. It is possible to show the order in which things happen on a collaboration diagram and even concepts such as iteration; however, these logical timings are very easy to get confused and can often detract from the model.

- Sequence diagrams have a far more rigid structure with regard to the layout of the model, as objects must be spread out across the page, whereas logical time must flow down the diagram. It is difficult to concentrate on interactions between groups of objects with a sequence diagram as the links often become very long and may cross several object lifelines before the reach their destination object. On the other hand, timing concepts such as iteration are far more visual using sequence diagrams.

Consider the chess example. The collaboration diagram gives a good indication of the interactions that may occur, generally, between the two players. However, if it was required to show a particular scenario (such as a simple checkmate known as fool's mate), this specific scenario would be far clearer if shown with a sequence diagram than if shown with a collaboration diagram.

Consider now the second example, that of the system life cycle. The collaboration diagram gives a good indication of the classical life cycle mode, in this case the Waterfall model. However, if it were required to model what actually happened on a project, then a sequence diagram would show the iterations far more clearly than the collaboration. If, on the other hand, a general overview of the life cycle model was required for training purposes, for example, the collaboration diagram would be the better of the two options.

5.6   Activity diagrams

5.6.1   Overview

This section looks at another behavioural model of a system: the activity diagram. Activity diagrams, generally, allow very low-level modelling to be performed compared to the behavioural models seen so far. Where interaction diagrams show the behaviour between objects, and statecharts show the behaviour within objects, activity diagrams may be used to model the behaviour within an operation, which is about as low as it is possible to go.

The other main use for activity diagrams and certainly the most commonly seen usage of activity diagrams in other texts is to model 'workflows'. This will be discussed in some detail later in this chapter as the level of abstraction of a workflow can vary.

Activity diagrams are actually a special type of statechart and, as will be seen, the constructs are very similar between the two types of diagram. Activity diagrams are also similar to traditional flow charts (from which they were derived), which is often a cause for comment or even criticism of the UML.

5.6.2   Diagram elements

Activity diagrams are made up of three basic elements: states (of which there are two types: action and activity), decisions and transitions. Already the similarity with statecharts can be seen.

The two types of state – action and activity – are subtly different. An activity state represents something that is happening that can be interrupted, whereas an action state represents a state that cannot be interrupted that is often referred to as being 'atomic'. Another difference between the two types of state is that activity states may be broken down into one or more further activity diagrams, which has the potential for a massive amount of nesting of models, which in turn can lead to much complexity. The two states may be differentiated graphically from normal states (as used in a statechart) as they are truly rounded boxes (they have no vertical line of the edges), whereas normal states are represented by boxes with rounded corners (vertical edges do exist).

Transitions are the same as transitions from statecharts in that they show how to cross from one state to another. Transitions may also have decisions placed upon them, which is represented graphically by a diamond symbol. This is, in some ways, an inappropriate choice of symbol for two reasons. The first reason is that the diamond symbol is already used to indicate aggregation, which may lead to some confusion when interpreting the diagrams. The second reason is that this is the point where the flow chart people start to complain that an activity diagram is little more than a glorified flow chart and, hence, should not exist!

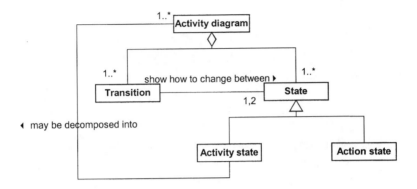

Figure 5.25 *Partial meta-model for activity diagrams*

It can be seen from Figure 5.25 that an 'Activity diagram' is made up of three or more 'State' (a start, end and at least one normal state – note that start and end states are not shown on this model) and two or more 'Transition' (to connect between the three states). A 'Transition' shows how to change between one or two 'State'. There are two types of 'State' – 'Activity' and 'Action' – and an 'Activity state' may be decomposed into an 'Activity diagram'.

Typically, an activity will be used to model an operation from a class diagram or activity from a statechart diagram.

5.6.3 Examples and modelling

The two main uses for activity diagrams are to model workflows and operations. Conceptually, modelling operations is far simpler to understand than modelling workflows. The first example, therefore, shows how to model the behaviour of an operation. This would be a procedure when applied to software or a general algorithm when applied to general systems (perhaps a thought process or hardware design).

The first example shows how to model a simple software algorithm that has been taken directly out of a programming algorithms book [17].

Figure 5.26 shows how a standard-defined algorithm for searching an array for a particular value may be modelled. The algorithm has been taken straight from a book on structured programming, which basically describes a library of useful algorithms that may be implemented in any programming language.

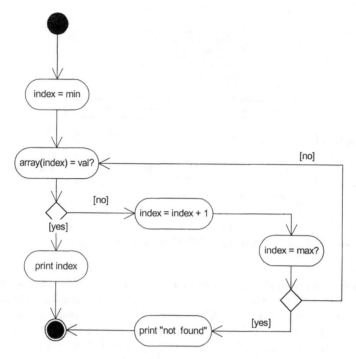

Figure 5.26 *Activity diagram describing the behaviour of an operation*

From the model, the first thing that happens is that 'index = min' where the attribute value (from an associated class diagram) is set to a predefined value of 'min'. Next, the array value indicated by 'index' is compared with 'val'. If the result of this comparison is 'yes' then 'print index' occurs, otherwise 'index = index + 1' occurs, where the 'index' attribute is incremented by one. Next 'index' is compared to a predefined value called 'max' to see if the end of the array has been reached. If not, the model reverts back to comparison state, otherwise the 'print 'not found'' state is entered.

Activity diagrams are very useful for modelling complex algorithms with much iteration, but may be 'overkill' for modelling very simple operations, particularly if many exist within the design. Indeed, some sort of simple functional descriptions, such as informal pseudo-code may be more appropriate for very simple operation definition.

The activity diagram can also be used to model any low-level conceptual operation that need not be software, as shown in the next model.

Figure 5.27 shows the 'move' operation from the original behavioural model lecture.

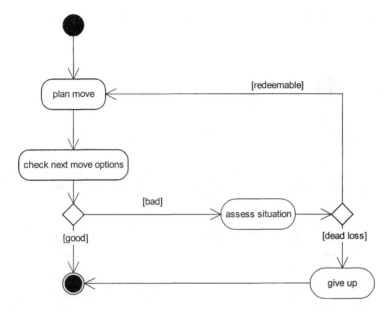

Figure 5.27 *Activity diagram showing behaviour of a thought process of how to play chess (badly)*

It can be seen that the first thing that happens is 'plan move'. The next is'check next move options', which is then followed by a decision branch. There are two outcomes: 'bad', which leads to 'assess situation', and 'good', which leads to the end of the activity diagram. Immediately following 'assess situation' is another decision branch where 'redeemable' loads back to 'plan move' and 'dead loss' leads to a 'give up' state.

The most widely used application (certainly from most other literature's point of view) for activity diagrams was concerned with modelling workflows. This can cause some confusion as the term 'workflow' is very –much associated with the Rational Unified Process (RUP) [3, 16], which has its own distinct terminology that can be different from some other terminology, such as the one adopted by ISO. Two examples of modelling workflows will be considered, one of an ISO-type interpretation of the term 'workflow' and one of the RUP interpretation of the term 'workflow'.

The first example will show a model of an ISO work practice, which is the lowest level of the ISO process structure and which is discussed in depth in Chapter 6.

Figure 5.28 shows how a change control procedure may be modelled as part of a process modelling exercise. This example is taken from a guide on how to implement ISO 9001, but rather than using text descriptions and flow charts as in the source reference [16], the procedure has been modelled using an activity diagram.

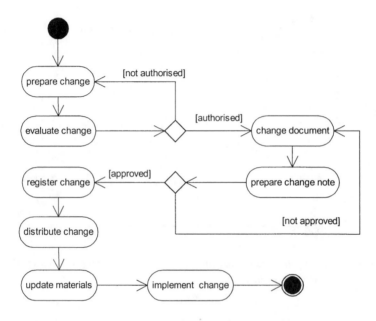

Figure 5.28 *Activity diagram for an ISO work practice*

From the model, the first thing that happens is 'prepare change', which is then immediately followed by an 'evaluate change' state. After this state, there is a decision box where the result of the previous state is 'authorised', before the procedure moves on to 'change document' and then 'prepare change note'. If the result is 'not authorised', the procedure reverts back to 'prepare change'.

After 'prepare change note' there is another decision box. If the result is 'approved', the procedure carries out the remaining states in series: 'register change', 'distribute change', 'update manuals' and 'implement change'. If the result is 'not approved', the procedure reverts back to the 'change document' state.

This model is directly compatible with the process structures that are discussed in Chapter 7.

The second example of modelling workflows is one that is taken from the definition of the RUP, which can be found in almost every UML textbook in existence today. For a more in-depth discussion of this concept of workflows, see

[2, 16]. For now, however, it is enough to know that the activity diagram is recommended by the RUP to be the diagram that is used to model workflows.

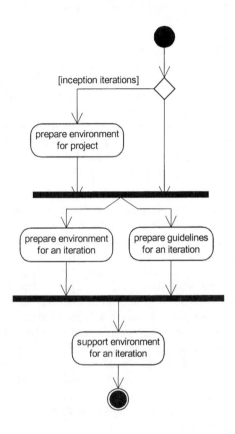

Figure 5.29 *Activity diagram showing workflow from the RUP*

Figure 5.29 shows an activity diagram that describes the behaviour of, or how to execute, a particular workflow. From the diagram, the first thing that happens is a decision that decides whether the workflow is in the 'inception iterations' or not. If the decision is true, the activity 'prepare environment for project' is executed; if not, the control is split using a fork. This splits the control so that parallel activities may be executed; in this case the activities that are executed in parallel are 'prepare environment for an iteration' and 'prepare guidelines for an iteration'. The control is then joined using a join that leads to the final activity 'support environment for an iteration'.

One point that should be stressed is that the states shown here are not necessarily from a single object as activity diagrams focus on the activity flow in a system-, rather than state-led view, as used in statecharts.

5.6.4   Using activity diagrams

Activity diagrams are one of the least used of the UML diagrams, which is reflected in the fact that many tools do not support activity diagrams! Perhaps one of the reasons why they are not used is because they are useful at the algorithm level. Unfortunately, in many cases and especially for particularly simple algorithms, creating activity diagrams is often perceived as a waste of time. Another reason why activity diagrams are underused is that there is some confusion about the exact nature and structure of the RUP, mainly due to the fact that the RUP is an evolving (yet well-proven) process.

Activity diagrams are actually defined as a special type of statechart, but it is sometimes difficult to see why two types of diagram are required when, it may be argued, they are very similar indeed. The most fundamental conceptual difference between activity diagrams and statecharts is that activity diagrams concentrate on activity flow and may thus show behaviour from more than one type of object, whereas a statechart only shows the behaviour from a single type of object. As a consequence, statecharts may show wait, or idle, states, whereas an activity diagram may not. The states in a statechart are normal states, which means that they may be complex and contain more than one internal transition, represented by a number of actions or activities, whereas activity and action states may only contain one. Activity diagrams may also use advanced syntax such as 'swimlanes' and 'object flow', which will be discussed briefly in Chapter 6. Swimlanes allow particular states to be grouped together and associated with a particular class or object, which is useful for assigning responsibility. Object flows allow the creation and destruction of objects or classes to be associated with states, which may be used to show data flow.

There is also a dangerous aspect to using activity diagrams due to the fact that activity states can be broken down into more, lower-level activity diagrams. This can be dangerous as it is very easy to get carried away and have many levels of deep-nested activity diagrams, which may hide many levels of complexity in a system.

One curious aspect of activity diagrams that is often commented upon is the shape of the decision element. It has already been mentioned that this leads to a small amount of confusion due to the fact that the decision tree symbol is also the same, graphically, as the aggregation symbol and that it lends comparisons to flow charts. The reason why this shape was chosen is that multiple decisions, such as 'switch' statements, may be split up, one branch at a time, so that they are clearer to see in the diagram. Imagine trying to show many transitions leaving a single state and what a mess the diagram would appear to be, compared to how it could be tidied up using the decision element.

5.7 Use case diagrams

5.7.1 Overview

This section introduces use case diagrams, which realise a behavioural view of the system. The behavioural view has an emphasis on functionality, rather than the control and logical timing of the system. The use case diagram represents the highest level of abstraction of a view that is available in the UML and it is used, primarily, to model requirements and contexts of a system. Use cases are covered in greater depth in Chapter 7 and thus this section is kept deliberately short, emphasising, as it does, the structure of the diagrams.

Use case diagrams are arguably the easiest diagram to get wrong in the UML. There are a number of reasons for this:

- The diagrams themselves look very simple, so simple in fact that they are often viewed as being a waste of time.

- It is very easy to go into too much detail on a use case model and to accidentally start analyses or design, rather than very high-level requirements and context modelling.

- Use case diagrams are very easy to confuse with data flow diagrams as they are often perceived as being similar. This is because the symbols look the same as both use cases (in use case diagrams) and processes (in a data flow diagram) are also represented by ellipses. In addition, both use cases and processes can be decomposed into lower-level elements.

- Information on the practical use of use cases is surprisingly sparse, bearing in mind that many approaches that are advocated using the UML are use case-driven. In addition, much of the literature concerning use case diagrams is either incorrect or has major parts of the diagrams missing!

With these points in mind, the remainder of this section will describe the mechanics of use case diagrams. For an in-depth discussion concerning the actual use of use case diagrams, see Chapter 6.

5.7.2 Diagram elements

Use case diagrams are composed of four basic elements: use cases, actors, associations and a system boundary. Each use case describes a requirement that exists in the system. These requirements may be related to other requirements or actors using associations. An association describes a very high-level relationship between two diagram elements that represents some sort of communication between them.

An actor represents the role of somebody or something that somehow interacts with the system.

The system boundary is used when describing the context of a system. Many texts and tools ignore the use of the system boundary or say that it is optional without really explaining why. Think of the system boundary as the context of the system and its use becomes quite clear and very useful. System boundaries and contexts are discussed in more detail in Chapter 6.

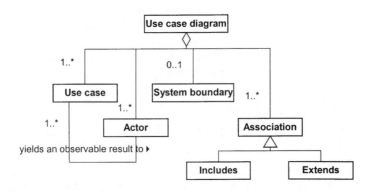

Figure 5.30 *Partial meta-model for use case diagrams*

Figure 5.30 shows that a 'Use case diagram' is made up of one or more 'Actor', one or more 'Use case', one or more 'Association' and zero or one 'System boundary'. A 'Use case' yields a result to an 'Actor'.

There are two special types of 'Association' (which have been defined as stereotypes, but are included here as they are included as a standard part of the UML): 'Includes' and 'Extends'.

The relationship between use case diagrams and other diagrams is not quite as clear-cut as with the other diagrams, which often leads to confusion. In order to explain this, consider the model in Figure 5.31.

Figure 5.31 shows that one or more 'Interaction diagram' shows an instance of a 'Use case'. An 'Interaction diagram' also shows interactions between one or more 'Objects', the behaviour of which may be described by a 'Statechart'. However, a 'Statechart' may also show the internal operation of a 'Use case'.

The source of ambiguity in the diagram is with the relationship between the statechart and the use case, and the statechart and the object. Statecharts have been discussed in some detail, particularly with regard to how they are used to model the behaviour of objects. However, statecharts can also be used to model the internal behaviour of a use case, which is at a higher level of abstraction than the object. Consider also, that an interaction diagram describes an instance of a use case, which can be interpreted as the statechart and interaction diagram being at the same level of abstraction, which goes against everything else that has been said so far concerning levels of abstraction of different diagrams. This is discussed in more detail in Chapter 7.

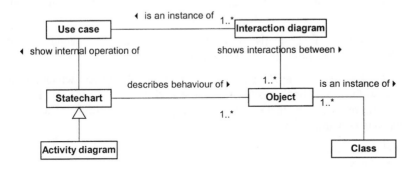

Figure 5.31 *Meta-model showing relationships between use cases and other UML elements*

5.7.3 Examples and modelling

Use case diagrams have two main uses: modelling contexts and modelling requirements. Although these two types of use should be related, they are distinctly different. For a full description of using use case diagrams, see Chapter 7. This section will concentrate purely on the mechanics and practical use of use case diagrams.

In addition, there are two types of context that can be modelled with regard to systems engineering: the system context and the business context. The first two example models show a business context and a system context for the ongoing chess game example.

Figure 5.32 shows the business context for the chess game. The business context shows the business requirements of the organisation and identifies actors (see the discussion on stakeholders in Chapter 7) that are associated with the business requirements. The business requirements that have been identified are as follows:

- 'make money', which is a business requirement that will be present on almost all business contexts in the world. This may seem obvious, but when asking for funding for a project the response from management is, invariably, 'make a business case and then we'll look at it'. By identifying the business requirements of your organisation, it is possible to start to justify expenditure for future projects.

- 'ensure quality' is a non-functional requirement that will impact on everything that the organisation does. This is particularly true when organisations are trying to obtain, or have already obtained, some form of standard accreditation.

- 'make games' is basically what the organisation does, which has two subtypes: 'make chess' and 'make draughts'.

- 'make chess' is the main business requirement that the system which has been used so far in this book has been concerned with. We can now see, however, how the requirement 'make chess' will help the organisation 'make money', which is very important.

- 'make draughts' is shown simply to indicate that the company may make more than one type of game.

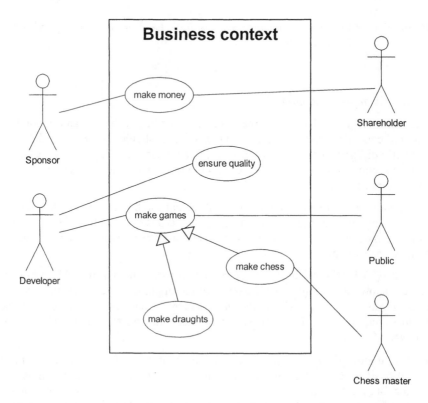

Figure 5.32 *Use case diagram showing the business context for the chess game*

The actors that have been identified for the system must each have some sort of association with at least one use case. If an actor exists which does not have an association with a use case, there is something seriously wrong with the model.

Figure 5.33 shows the system context, rather than the business context. It is important that these two contexts can be related together in order to justify why a particular project should exist.

Figure 5.33 shows the system context for the chess game. The use cases that have been identified are, at this level, very simple and very high level. The use case 'play chess' represents the main use of the chess system that is being developed. The second use case, 'teach chess', shows the secondary function of the chess system, which is to teach a 'Learner' how to play chess. One argument that is often levelled at use case diagrams is that they are too simple and fail to add anything to a project. However, no matter how simple, a good use case diagram can contribute to a project in several ways:

- They can be traced back to a business context to help justify a project. Quite often it is the case that a simple one-sentence description is required to sum up a whole project and a good system context will show exactly that.

- They can give a very simple overview of exactly what the project is intending to achieve and for whom.

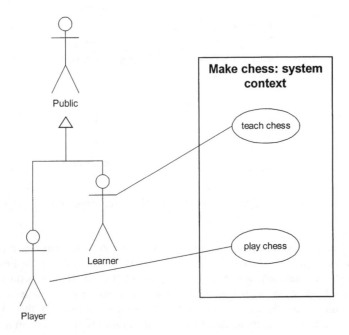

Figure 5.33 Use case diagram showing system context for the chess game

- They can form the basis of lower-level, decomposed use case diagrams that add more detail.

It is also worth remembering that just because a use case diagram looks simple, it does not mean that it took somebody two minutes to create and did not involve much effort. This is only true when it comes to realising the model using a CASE tool, but even the most simple of diagrams may take many hours, days or even months of work. Remember that what is being visualised here is the result of requirements capture, rather than the actual work itself.

Figure 5.34 shows how a single requirement may be taken and modelled in more detail, by decomposing the requirement into lower-level requirements.

Figure 5.34 shows the 'play chess' requirement that has been decomposed into three lower-level requirements: 'read set up', 'indicate move' and 'show thinking'. Each of these requirements is a component requirement of the higher-level requirement, which is indicated by the special type of association '<<includes>>', which shows an aggregation-style relationship.

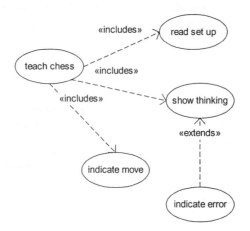

Figure 5.34 *Use case diagram showing modelling requirements*

The second special type of association is also shown here and is represented by the stereotype '<<extends>>'. The '<<extends>>' relationship implies that the associated use case somehow changes the functionality of what goes on inside the higher-level use case. In the example here, the extending use case is 'indicate error', which extends the functionality of 'indicate move' in the event that something untoward happens and forces 'indicate move' to behave in a different way.

Figure 5.35 shows what a use case diagram should not look like, which is a mistake that is very easy to make.

The model in Figure 5.35 shows how a model should not be created as it defies the objectives of use case diagrams. There are several points to consider here:

- The definition of a use case is that it must yield some observable result to an actor. In the example shown here, the decomposed use cases do not yield any observable result to the actors – they are modelled at too low a level.

- Remember that the aim of modelling requirements is to state a user's needs at a high level and not to constrain the user with proposed solutions. The example shown here has done exactly that and has started down the design road – which is inappropriate at this level.

- The diagram is turning into a data flow diagram by starting at the highest level and decomposing until the problem is solved. Use cases are not the same as data flow diagrams, despite appearances, and should not be used as such.

These points are discussed in more detail in Chapter 7.

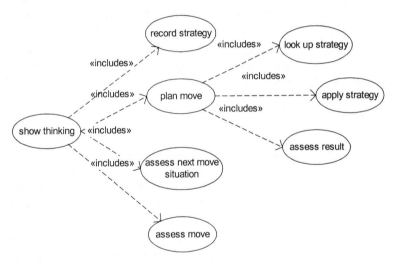

Figure 5.35 *Use case diagram showing a poor example of modelling*

5.7.4 Using use case diagrams

Use case diagrams are, visually, one of the simplest diagrams in the UML, but they are also the most difficult to get right. Part of their downfall is this simplicity, which makes them easy to dismiss and easy to get wrong. Another problem with use case diagrams is that they look like data flow diagrams and thus people tend to over-decompose them and start carrying out analysis and design rather than just simple requirements modelling.

One of the strengths of use case diagrams only comes out when they are compared with contexts. It is important to be able to trace any requirement back to the business context to justify a business case for the project. It is also important to be able to trace to the system context as this provides the starting point for the project.

The meta-model shown here represents all of the basic UML elements of use case diagrams, which indicates the simplicity of the diagram.

5.8   Component diagrams

5.8.1   Overview

This section introduces component diagrams that realise a static view of a system. Component diagrams are used, in the main, at later stages of system development when it comes to packaging up the final product and delivering to the customer. Component diagrams show real-life aspects of a system and have strong relationships to both class and object diagrams.

The simplest way to summarise why a component diagram is used is to imagine buying any product that is packaged in a box. The component diagram, basically, shows 'what's in the box'.

5.8.2   Diagram elements

Component diagrams are made up of several basic elements: components, interfaces and relationships. Components represent pre-existing entities that you may find 'in the box' and that may need to be assembled or configured before use. Interfaces represent the functionality of components that is directly available to the user. Relationships, as the name implies, represent conceptual relationships between such components.

Figure 5.36 shows the partial meta-model for component diagrams. From the diagram, a 'Component diagram' is made up of one or more 'Component', zero or more 'Relationship' and zero or more 'Interface'.

'Relationships' relate together two 'Component' and each 'Interface' shows the usable functionality of a 'Component', which is available to the user of the component.

One aspect of component diagrams that is often unclear is how, exactly, components relate to the rest of the UML model of a system.

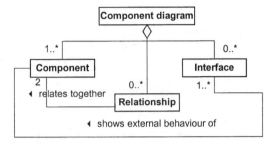

Figure 5.36 *Partial meta-model for component diagrams*

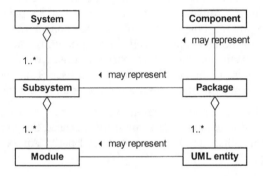

Figure 5.37 *Systems and components*

Figure 5.37 shows how components relate to a system. Any generic system may be thought of as being made up of one or more subsystems, each of which is made up of one or more modules. This may be related to UML constructs, such as a 'Package' being made up of one or more 'UML entity', as seen in Chapter 3. A 'UML entity' may represent a 'Module' and a 'Package' may represent a 'Subsystem'. Think of the 'Component' as a formal representation of a 'Package'. By tracing this relationship one step further, it tells us that a 'Component' is strongly related to a collection of UML entities, or a subsystem.

5.8.3 Examples and modelling

One of the main uses for component diagrams is to show components and their interfaces. Curiously, this is often hailed as a great use, but interfaces are, in reality, rarely used. Figure 5.38 shows a component and two simple interfaces.

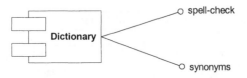

Figure 5.38 *Example component and interfaces*

Figure 5.38 shows an example component diagram showing a single component, represented graphically by a rectangle with two smaller rectangles on top of it (a legacy of a previous Booch symbol), with two interfaces that are shown graphically by a line that terminates with a small circle.

The component in this case, 'Dictionary', is represented graphically by a large rectangle with two smaller rectangles crossing one side. Two interfaces are shown – 'spell-check' and 'synonyms' – that are represented graphically by circles. Interfaces here represent functionality of the component that is available to anything external to the component itself. In simple terms, these two interfaces would relate directly to operations that exist on classes within the component, which would otherwise be hidden from the system user. This component would then be connected to other components via such compatible interfaces.

A typical use for a component diagram would be to show what would actually be delivered to a customer or user. In some software cases, this may be a model of the source code files that may be given to a third party. An example of this is shown in Figure 5.39.

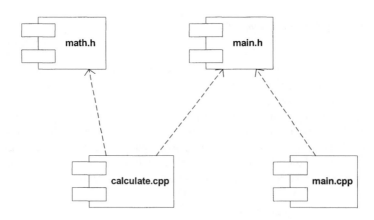

Figure 5.39 *Component diagram showing how to model source code*

Figure 5.39 shows the components that represent source code files with no interfaces [2]. The files shown include actual source code ('calculate.cpp' and 'main.cpp') along with their associated, header files ('math.h' and 'main.h').

Dependencies between these files are shown by directed lines. These dependencies, in this case, represent compilation dependencies.

This may be taken one step further in the case of software by showing the actual delivered executable files that are delivered to a customer, as shown in Figure 5.40.

Figure 5.40 shows an example of an executable release, which will include the executable versions of the source code [2]. However, in many cases, the software relies upon other parts that are needed to run the software. These other parts many include '.dll' files, '.class' or '.jav' files, etc. By showing these as components, it is possible to show configuration of executable releases for a particular piece of software. The diagram here also shows version numbers of the

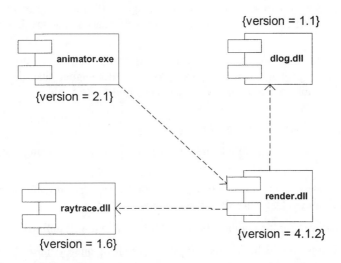

Figure 5.40 *Component diagram showing an executable release*

software and files, shown as 'tagged values', which is one of the UML extension mechanisms. Tagged values are discussed in more detail in Chapter 8.

Component diagrams, like all other UML diagrams, do not necessarily have to be related to software systems and can be related to anything. In order to demonstrate this, we revisit the chess example and show a component diagram that has been created.

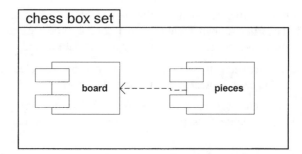

Figure 5.41 *Component diagram showing how to model a physical system*

The model in Figure 5.41 shows two components – 'board' and 'pieces' – with a single dependency between them. Relating these components to other UML elements, one would expect to find two packages with the same name. In this case, it is not difficult to imagine two groups of classes that are held in two packages.

Also on this diagram, it can be seen that the two components are collected together into a package named 'chess box set'. Remember that the package represents a simple grouping of UML elements, almost like a folder.

5.8.4   Using component diagrams

There are a few rules of thumb that should be applied when using component diagrams:

- Physical aspects of the system are modelled that relate directly to real-life physical things. These are known as concrete instances.

- The interfaces for the component should be small and well defined.

- Components should be a direct implementation of one or more classes that should already exist in other diagrams, or packages.

Component diagrams are typically used towards the end of a project life cycle as they show 'what's in the box' when a product is delivered. This will include source code and executable code in the case of software, but, basically, will represent anything that is delivered to the customer. However, component diagrams can be used at any point in the life cycle and some people even use them at the requirements phase to divide the main system into subsystems.

5.9 Deployment diagrams

5.9.1 Overview

This section introduces deployment diagrams, which realise a static view of the system. Like component diagrams, deployment diagrams are often used towards the end of a project life cycle when the system has been packaged up. If a component diagram shows 'what's in the box', it is the deployment diagram that shows 'where to put the pieces'.

5.9.2 Diagram elements

Deployment diagrams are made up of two main elements: nodes and links. Nodes represent real-world aspects of a system and show where components from the component diagram are housed. Nodes are represented graphically by a 3-D box, upon which objects or components are shown. The objects and components come from their respective diagrams. In software terms, a node may represent a computational resource such as a processor or computer. Links show relationships between nodes and are represented graphically by a simple line.

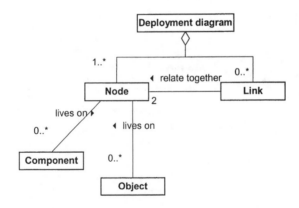

Figure 5.42 *Partial meta-model for deployment diagrams*

Figure 5.42 shows that a 'Deployment diagram' is made up of one or more 'Node' and zero or more 'Link'. A 'Link' relates together two 'Node'.

Zero or more 'Component' may live on a 'Node' and zero or more 'Object' may live on a 'Node'.

The objects are derived from any other diagram that includes objects, such as the object diagram and interaction diagrams (both collaboration and sequence diagrams). The components are derived, unsurprisingly, from component diagrams.

5.9.3　Examples and modelling

There are four main uses for deployment diagrams which are: modelling embedded systems, modelling distributed systems, modelling client server systems and modelling physical systems. An example diagram for each of these uses will be discussed in this section.

Figure 5.43 shows an embedded system that represents a robot arm controller system. The actual part of the system that has been the subject of the project is the controller software, which is represented here as the 'controller.exe' component that lives on the node 'controller' and that happens to be a '<<processor>>'. Other elements in the diagram include 'shoulder', 'elbow' and 'wrist', each of which happens to be a '<<motor>>'. The final element is the 'manual control', which happens to be a '<<keypad>>'.

Note the extensive use of stereotypes, which is typical for deployment diagrams. Stereotypes are discussed in more detail in Chapter 8 and caution must be exercised when using them.

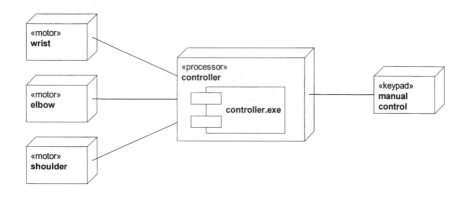

Figure 5.43 *Deployment diagram showing embedded systems*

Figure 5.44 shows how a distributed system may be modelled using a deployment diagram. The example chosen here is used as part of the user manual for a website in order to show the customer where the original website files must end up before the website will work and be accessible from the Internet.

Figure 5.44 shows a distributed system that has been modelled using a deployment diagram.

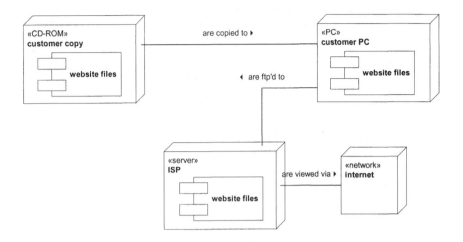

Figure 5.44 *Deployment diagram showing distributed systems*

The node on the left is called 'customer copy', which happens to be a '<<CD-ROM>>' that represents the actual disc that is delivered to the customer. The next node is called 'customer PC', which happens to be a '<<PC>>'. This represents the customer's computer on to which the information on the CD-ROM must be copied. This is apparent from the diagram as the directory that must be copied across to is represented by the component 'website files' and the link between the two nodes describes the nature and direction of the relationship.

The third node, called 'ISP' and which happens to be a '<<server>>', represents the customer's Internet service provider (ISP) to which the 'website' files must be ftp'd.

Note again the extensive use of stereotypes and remember that this diagram would lose a great deal of meaning without some sort of legend to explain what the stereotypes mean. Such a legend is known as the 'assumption model', which is discussed in detail in Chapter 8.

The next example of the use of a deployment diagram shows a client-server system that makes use of packages.

Figure 5.45 shows a simple client-server application that shows how a basic home-banking system works. The nodes used here represent the 'customer PC', which happens to be a '<<PC>>', the 'account query', which happens to be a '<<server>>', and the 'account details', which happens to be a '<<database>>'.

Notice how UML packages are introduced here to group the client components into one package and all server components into another package.

Without these packages, the diagram would lose a great deal of meaning and it would certainly not be immediately apparent that this was a client-server system.

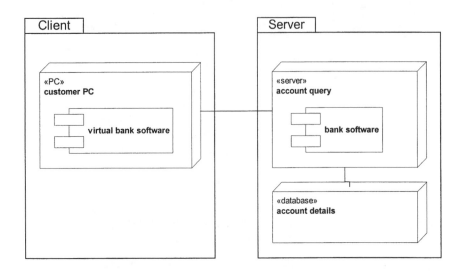

Figure 5.45 *Deployment diagram showing a client-server system*

The final example of the use of deployment diagrams pays another visit to the game of chess. This time, as we know 'what's in the box' from the component diagram, we need to know where to put the things that are in the box. This may seem like stating the obvious, but it wraps up the chess example neatly and goes to show that it is possible to model just about anything with the UML, including where to sit during a game of chess!

The components that were defined in Figure 5.41 showed that two components are available: 'pieces' and 'board'. We also know from the object diagram in Figure 5.11 and the interaction diagrams in Figure 5.19 and Figure 5.23 that two objects exist: 'Player 1' and 'Player 2'. The deployment diagram relates all these elements together to show where they are deployed.

Figure 5.46 shows how the components and objects defined in a chess game box set and the associated players are deployed in the real physical world.

There are two nodes that happen to be '<<chair>>', which are 'my chair' and 'your chair', where the objects 'Player 1' and 'Player 2' are deployed, or sit. The 'board' and 'pieces' from the component diagram are deployed on the 'kitchen table', which happens to be a '<<table>>>'.

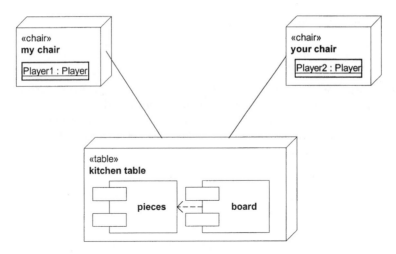

Figure 5.46 *Deployment diagram showing physical systems layout for the chess example*

5.9.4 Using deployment diagrams

Deployment diagrams, like component diagrams, have a very limited syntax and thus the meta-model shown here is almost complete. Advanced concepts that may be used for deployment diagrams include different types of nodes to represent processing nodes and non-processing nodes.

Deployment diagrams are also one of the more little-used UML diagrams, which is reinforced by the lack of CASE tool support for the diagram.

An interesting point about deployment diagrams is that, in most cases, they are used towards the end of a project when the final product is being packaged up; however, in some cases, they are the first diagram to be used. A particular application occurred where an existing system had to be modelled, in this case a telecommunications network, and it was perceived that the best approach for modelling the system was to start with the finished product and then to work backwards, or reverse engineer the system, to find out how it worked. This just goes to show that there are virtually no strict guidelines when it comes to how to apply the UML, providing that the models are all correct and consistent.

5.10 Summary and conclusions

5.10.1 Summary

This chapter has introduced each of the nine UML diagrams in turn and has provided examples of their use. Each diagram has been discussed at a very high level, often missing out much of the diagram syntax, so that the modelling aspects of each diagram could be focused on, rather than being bogged down by all the syntax of each diagram.

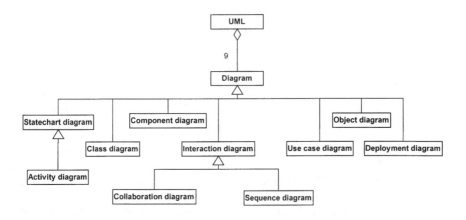

Figure 5.47 *The nine types of UML diagram*

The following models serve as a recap for everything that has been said in this chapter, which should be able to be read by anyone who has read (and understood) this book so far. Even with the limited amount of syntax that has been introduced, it is still possible to model many aspects of a system.

Figure 5.47 shows the nine types of UML diagram that have been introduced in this chapter. Note how the 'Sequence diagram' and 'Collaboration diagram' are grouped into a generalised class called 'Interaction diagram', which reflects that fact each of the two diagrams show the same sort of information. Note also how the 'Activity diagram' is a specialised type of 'Statechart diagram'.

These nine diagrams are often grouped into categories to represent different views of the system. Rather than be prescriptive with these groupings, the diagrams in this book are grouped into two very broad types of model: behavioural and static.

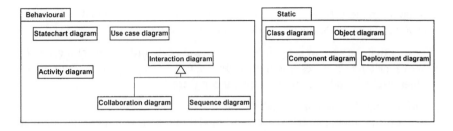

Figure 5.48 *The types of diagram grouped into types of model*

Figure 5.48 shows how the nine types of UML diagram are grouped into the two types of model. Many other texts on the market will show different 'views' of a system, such as logical view, physical view, use case view, etc, and this is fine but it will not fit every application. These views are also consistent with the two types of model shown here; each type of view that is defined may consist of a static and behavioural element. Treat the two models here as the absolute highest level of types of model and use the views introduced in other books as necessary.

5.11 Conclusions

In order to conclude this chapter, there are a few pieces of practical advice that should be borne in mind when modelling using the UML. These rules are not carved in stone, but experience has shown that they will be applicable to 99 per cent of all modelling.

- Use whatever diagrams are appropriate. There is nothing to say that all nine diagrams should be used in order to have a fully defined system – just use whatever diagrams are the most appropriate.

- Use whatever syntax is appropriate. The syntax introduced in this book represents only a fraction of the very rich UML language. It is possible to model most aspects of a system using the limited syntax introduced here. As you encounter situations that your known syntax cannot cope with, it is time to learn some more. There is a very good chance that there is a mechanism there, somewhere that will.

- Ensure consistency between models. One of the most powerful aspects of the UML is the ability to check the consistency between diagrams which is often glossed over. Certainly, in order to give a good level of confidence in your models, these consistency checks are essential.

- Iterate. Nobody ever gets a model right the first time, so iterate! A model is a living entity that will evolve over time and, as the model becomes more refined, so the connection to reality will draw close.

- Keep all models. Never throw away a model, even if it is deemed as incorrect, as it will help you to document decisions made as the design has evolved.

- Ensure that the system is modelled in both views. In order to meet most of the above criteria, it is essential that the system is modelled in both views, otherwise the model is, quite simply, incomplete.

- Ensure that the system is modelled at several levels of abstraction. This was one of the fundamental aspects of modelling and will help to maintain consistency checks, as demonstrated in the chess example.

Finally, modelling using the UML should not change the way that you work, but should aid communication and help to avoid ambiguities. Model as many things as possible, as often as possible, as the more that you use the UML, the more benefits you will discover.

5.12 Further discussion

1. Take one of the partial meta-models shown in this chapter and then select any other UML textbook – see the references at the end of this chapter for a list of alternative. Based on the detailed information in the textbook, populate the meta-model further to gain a greater understanding of the type of diagram chosen for this example.

2. Model a set of chess prices using a class diagram. How will the different types of pieces be represented – by colour or by movement?

3. Expand on the chess game example by adding a new operation called 'think' to the class diagram introduced in Figure 5.2. Which other diagrams will this affect?

4. Expand on the chess example in Figure 5.2 by defining two types of 'Player': 'Human' and 'Computer'. What impact will this have on the model and which models will need to be redefined or created from scratch?

5. Take any pre-recorded CD-ROM and model the contents, based on what can be seen, for example, from a typical Windows-style browser.

6. Now model the steps that you took in order to put the CD into the computer and read the contents of the CD-ROM. Think about deployment and component diagrams for the static view and interaction diagrams and statecharts to model how this was achieved.

5.13 References

1   RUMBAUGH, J., JACOBSON, I. and BOOCH, G.: 'The unified modelling language reference manual' (Addison Wesley, 1998)

2   BOOCH, G., RUMBAUGH, J. and JACOBSON, I.: 'The unified modelling language user guide' (Addison Wesley, 1998)

3   JACOBSON, I., BOOCH, G. and RUMBAUGH, J.:, 'The unified software development process' (Addison Wesley, 1999)

4   DOUGLASS, B. P.: 'Real-time UML' (Addison Wesley, 1998)

5   FOWLER, M. with SCOTT, K.: 'UML distilled (applying the standard object modelling language)' (Addison Wesley, 1997)

6   LEE, R. C. and TEPFENHART, W. M.: 'UML and C++, a practical guide to object-oriented development' (Prentice Hall, 1997)

7   LARMAN, G.: 'Applying UML and patterns – an introduction to object-oriented analysis and design' (Prentice Hall Inc., 1998)

8   ROYCE, W.: 'Software Project Management – A unified framework' (Addison Wesley, 1998)

9   CANTOR, M. R.: 'Object-oriented project management with UML' (John Wiley and Sons Inc., 1998)

10 SI ALHIR, S.: 'UML in a nutshell – a desktop quick reference' (O'Reilly and Associates Inc., 1998)

11 ERIKSSON, H. and PENKER, M.: 'UML Toolkit' (John Wiley and Sons Inc., 1998)

12 MULLER, P.: 'Instant UML' (Wrox Press Ltd, 1997)

13 STEVENS, P. and POOLEY, R.: 'Using UML' (Addison Wesley, 1999)

14 SCHMULLER, J.: 'SAMS teach yourself UML in 24 hours' (SAMS, 1999)

15 HOYLE, D.: 'ISO 9000, pocket guide' (Butterworth Heinemann, 1998)

16 KRUCHTEN, P.: 'The Rational Unified Process: an introduction' (Addison-Wesley Publishing, 1999)

17 RADER, J. R.: 'Advanced software design techniques' (Petrocelli Books, 1978)

PART III: Solutions

Chapter 6

Modelling standards, processes and procedures

"there ain't no rules around here, we're trying to accomplish
something"
– Thomas Edison

6.1 Introduction

6.1.1 Introduction

In this chapter, standards, processes and procedures are modelled using the
Unified Modelling Language (UML). The importance of having a defined process
has been discussed previously in Part I. This chapter looks at how the UML may
be used to be effectively model processes in a correct and unambiguous fashion in
order to minimise complexity of the process and to maximise the effective
communication involved with implementing the process.

This section is split into six main sections:

- Introduction, where the rationale for modelling standards, processes and
  procedures is discussed.

- Analysing standards, where the UML is used to model and help to understand
  standards.

- Defining the procedure, where the analysis models from the previous sections
  are used as a reference for defining a new procedure.

- Life cycles, where life cycles and life cycle models will be modelled and then
  related to the new process defined in the previous section. In addition, the
  process will be tailored to allow for projects with unique requirements.

- Implementing the process, where the structures of the output from the newly
  defined process are modelled. Practical issues for implementing the procedure
  will be addressed, which will form the basis of using a systems engineering
  tool.

- Finally, conclusions are drawn about the whole issue of modelling standards.
  For other examples of process modelling, see [1].

6.2   Standards

The importance of having a well-defined process was discussed in Chapter 1. However, it is not good enough simply to have a defined process, as it is also necessary to demonstrate compliance with a norm, demonstrate usage of the process on projects and demonstrate that the process is effective. It is possible to use the UML in order to achieve these aims, and this chapter sets out to do exactly that.

The first of the points raised here, demonstrating that the process is compliant with an existing norm, is crucial to allow the process to be audited or assessed according to such a norm. In real terms, these norms are standards, procedures, best practice guides, textbooks [2], etc.

This may seem perfectly reasonable and fit in with our definition of systems engineering being the implementation of common sense, but this is far easier said than done. In order to demonstrate compliance with a standard it is important to have a good understanding of that standard. In addition, it is necessary to be able to map between the relevant part of the chosen standard and the process under assessment.

6.2.1   Standards, standards, standards

Unfortunately, there are many problems associated with standards that hinder their understanding, or that lead to incorrect implementation of standards, and that were introduced briefly previously in this book.

One problem with standards is that there are so many of them – in fact, in some ways there are too many standards, which may lead to all sorts of problems. Here is a list of some of the more popular standards in use today in the world of systems engineering:

- **EIA 632 Processes for engineering a system**. This is a popular standard that has seen a lot of implementation in the defence and aerospace industries.

- **EIA 731 Systems engineering capability model**. This is a widely adopted standard that is concerned with capability determination and uses the processes defined in EIA 632 Processes for engineering a system.

- **ISO 9001 Model for quality assurance in design, development, production, installation and servicing**. This is arguably the most widely used standard in the world as it provides the international benchmark for system quality. This includes all aspects of systems except for software-based systems! This has led to the development of a set of guidelines specifically for software systems, which is described in the next point.

- **ISO 9000-3 Guidelines for the application of 9001 to the development supply and maintenance of software**. This is not a standard but a set of guidelines that describe how to apply ISO 9001 to software systems. This has also led to the definition of formal accreditation guidelines under the TickIt scheme.

- **IEC 61508 Functional safety of electrical/electronic/programmable electronic safety-related systems**. This standard is aimed very much at programmable devices, such as PLCs, and is used heavily in the signalling world.

- **ISO 15288 Life cycle management – system life cycle processes**. The latest attempt by ISO to make sense of the standards maze that has been generated in conjunction with major industry, which promises a large uptake in many industries [3].

- **ISO 12207 Software life cycle processes**. This standard provides an input to many of the other standards mentioned here and is one of the most widely accepted standards. In fact, most standards will make explicit reference to ISO 12207 and cite it as a basis for life cycle definition.

- **ISO 15504 Software process improvement and capability determination (SPICE)**. This very long and well-written standard defines a complete set of processes within an organisation that can be used as the basis for process improvement and/or capability determination.

- **IEEE 1220 Application and management of the systems engineering process**. This US standard is a widely used international standard that has a strong track record for systems development.

These are just a small selection of standards that exist in the world of systems and do not include any industry-specific standards. Anybody familiar with standards should recognise at least one of those in the list above. Anybody who has ever tried to read and understand any of these standards (or any standard, for that matter) may have encountered some difficulty in understanding them, which may be for any number of reasons:

- Standards are often very difficult to understand individually and, it may be argued, some are not terribly well written. As will be demonstrated later in this chapter, some standards are inconsistent with themselves and contain many errors and much ambiguity. This does not help when trying to read and understand standards (a lack of understanding), which puts us on the rocky road to complexity and communications problems. If a standard is not fully understood, then this may impact other processes, such as training and assurance etc.

- If standards are difficult to understand individually, they are almost impossible to understand when they are read in groups. It is often the case that an organisation wants to comply with a generic standard, such as ISO 9001, but also wants to comply with a more industry-specific standard (e.g. EN 50128 Railway applications, software for railway control and protection systems) or even a more field-specific standard (e.g. ISO 15288 Life cycle management – system life cycle processes).

- Understanding the information may also be hindered by the fact that some standards are too long (some in excess of 300 pages), which can seem insurmountable when first encountered. Length is not necessarily an indication of complexity but in many cases will lead to such. Indeed, some of the longer standards are actually less complex than those with a far lower page count. For example, ISO 15504 is far less open to ambiguity that ISO 9001, despite having a page count that is over ten times higher.

- Some standards are too short and written at such a high level that they bear little or no resemblance to what actually occurs in real life. Take, for example, ISO 9001, which, it can be argued, is the most widely used standard in the world today. It is perhaps because it is applicable to so many different areas that the standard is so generic. However, there is an inherent danger in this, as the more generic the standard, the more open to ambiguity it becomes. These ambiguities occur here as two different people reading the same paragraph of the standard may come up with two completely different interpretations.

- Standards are often written by committees and thus tend to end up somewhat less than comprehensible. Standards often take years to be prepared and trying to keep all of the people happy all of the time can often lead to the production of a camel rather than a horse!

Assuming then, for one moment, that the standards can be understood, the next issue is that of demonstrating compliance. Compliance is demonstrated through audits or assessments. Audits are performed by an external, registered organisation and yield a straight pass or fail result. Assessments, on the other hand, may be carried out internally or externally, but usually yield more useful results, such as a capability profile. As a simple example of this, ISO 9001 requires a formal audit, which yields a 'yes or no' outcome. ISO 15504, on the other hand, requires an assessment (as opposed to an audit) that yields a potentially more useful capability profile that can be used to improve existing processes.

Compliance is actually achieved through having a defined process that is then the subject of an audit or assessment. In almost all cases, standards explicitly state that a defined process is essential to demonstrate compliance.

Most audits and assessments will call for some sort of mapping to be established between the process and the standard that is being used as a basis for the assessment or audits. In order to map between two of anything, it is important that they are both represented in a language or format that may be directly compared – otherwise mapping is impossible.

This then means that if any of these standards are to be met, they must be able to be related to a defined process. In addition, if more than a single standard is read, they must be able to be compared and contrasted so that a common understanding may be achieved.

6.2.2 Requirements for using standards

In summary, therefore, it is possible to draw up a set of requirements that are needed in order to use standards in a practical and effective manner:

- There is a need to be able to understand standards. This means that any problem areas such as ambiguities or inconsistencies can be sorted out and that everyone who reads the standard has a common understanding.

- It is vital to be able to compare standards with a defined process. It is no use at all if both the standard and the process can be understood individually if they cannot be compared in an effective manner.

- It is also necessary to be able to compare different standards with one another. Again, standards often refer to other standards, or are required to be met in conjunction with some other standard.

It is essential, therefore, that all standards and processes can be represented in a common format or language. The UML represents such a common language and, as will be demonstrated, is most effective at modelling processes and standards. This all boils down to being able to effectively analyse any standard or process by modelling them using the UML.

6.3 Analysing standards

Analysing may be thought of as understanding the problem at hand. In this case, the problem at hand is the actual standard, or standards, that need to be modelled.

Analysis may be performed practically by modelling information contained within a standard, or to put it another way, to model the actual content of a standard using the UML. Once models have been created in the UML, they may be compared and contrasted together in order to gain a greater understanding of them. It is then possible to use these models as a basis for modelling a new standard that may be completely modelled and designed before it is actually written.

Another benefit of modelling is that potential areas of complexity may be highlighted and either eliminated or addressed in some other way. All these points will be demonstrated in this chapter.

6.3.1 Aspects of modelling using the UML

It has been stated many times that the UML may be used to model standards, but which of the nine diagrams that are available are suitable and what aspects of the standards may be modelled? The simple answer to this question is to use whatever diagrams seem appropriate, but this answer does not help us much in a practical way. Therefore, some UML diagrams and their usage are suggested here, but these are by no means carved in stone:

- Class diagrams, which form the backbone of the UML, may be used to model almost all static aspects of a standard or process. This includes modelling the initial high-level views of a standard that define the way in which the whole process will be laid out. Class diagrams may also be used to show the types of process, task, activity, or whatever nomenclature is being adopted. This may also be taken a step further by modelling actual deliverables associated with particular processes. This will not only show the deliverable and which process it may be associated with, but will also show the actual structure of the deliverable, which may be used as the basis for a template. Implementing a process may also be made easier by defining an information model and implementation model that can then be the basis for implementation using an automated tool.

- The behavioural aspects of processes – that is to say in what order things must be done and under what conditions – may be shown by using statecharts or activity diagrams that are associated with classes that represent processes. This can be particularly useful from a project management point of view and also for training staff in how they should be carrying out a particular process.

- At a higher level, interaction diagrams may be used to model life cycle models that show both how the project is to be carried out (using collaboration diagrams) and how the project is actually carried out (using sequence diagrams).

- At a very low level, activity diagrams may be used to state explicitly how particular process steps are to be executed for a very rigidly defined process.

All of these techniques have been shown to be very effective for process modelling for both analysis and design of processes. The first step that will be covered here, however, will be how to analyse two standards.

6.3.2 Modelling example

Imagine a situation where it was required to produce a new process for an organisation that needed to meet a particular systems engineering standard and also generic ISO processes. In the example here, the systems engineering standard that has been chosen is EIA 632 – Processes for engineering a system, and the ISO processes are to be based on those defined in ISO 15504 – Software process improvement and capability determination (SPICE)'.

This is not such an unusual combination as it may first appear, as ISO 15504 is a very well-specified standard and is fully compliant with ISO 9001. One of the advantages of ISO 15504 over ISO 9001, however, is that it is far more rigorously defined and is thus less open to ambiguity than ISO 9001. Showing any pre-worked example in a book can have its own danger. The danger with using such an example, however, is that it can appear to be like the work of a magician, where words and diagrams may be pulled, like rabbits, from a hat, yet nothing much will be said or learned – but it certainly looks good! To avoid such a display of sleight of hand, several other international standards and processes will be presented towards the end of this section along with a discussion on how they could be related to the specific standard models shown here.

Having established the groundwork, it is time to dive straight in and start to model the standards.

6.3.3   EIA 632 Processes for engineering a system

The first model that will be drawn up is what will be referred the high-level static model. There is no rule that says that this must always be the first model, but experience has shown that it is a useful one to get out of the way early on in the modelling.

The high-level structural model really sets the scene for the remainder of the modelling and contains the most fundamental concepts of the standard itself. It is crucial that this model can be clearly and unambiguously defined, otherwise the rest of the standard is rendered meaningless. The idea is to try to extract the actual structure, or hierarchy, of the processes that are defined in the standard. This is achieved by looking for UML-friendly words in the process descriptions in the standard, such as: 'is a type of', 'is made up of', etc.

Figure 6.1 shows the high-level structural model that was extracted for EIA 632.

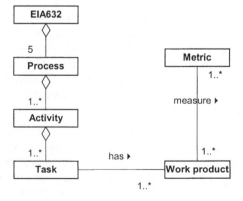

Figure 6.1   *High-level structure of EIA 632*

Figure 6.1 shows the high-level structure of EIA 632 and is read as follows.

'EIA 632' is made up of five 'Process'. This is the highest-level piece of information that could be extracted and states the whole structure of the standard itself. This information is quite easy to extract as the five processes form the major sections in the standard.

The next item of information is the structure of 'Process', which is made up of one or more 'Activity', each of which is made up of one or more 'Task'. Each 'Task' has one or more 'Work product' associated with it. One or more 'Metric' measures one or more 'Work product'. This information was also quite easy to extract from the standard and is stated explicitly in the standard itself.

In a nutshell, this model describes the whole structure of the standard. All subsequent models will go into greater detail, but will have to be consistent with this model.

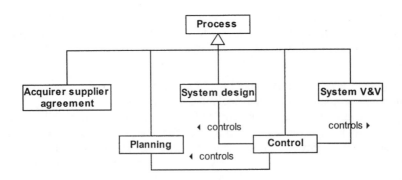

Figure 6.2   *Types of process in EIA 632*

The next step is to concentrate on one of the classes – in this case, 'Process'. It was stated in the previous model that there are five processes. Remember that the term 'is a type of' is a term used in the UML, which provides the basis of the model shown in Figure 6.2. Therefore, the diagram reads: there are five types of 'Process', which are 'Acquirer supplier agreement', 'Planning', 'System design', 'Control' and 'System V&V'. Several other relationships also exist in the model, which are: 'Control' controls, 'Planning', 'System design' and 'System V&V'.

As an interesting aside, note how 'Control' does not control 'Acquirer supplier agreement'. Is this correct, or is it an oversight by the authors? Already, even at this high level, it is possible to throw up potential inconsistencies or problems with the standard. This is an issue that may have to be addressed before the modelling can be taken any further and the full implications of the lack of 'Control' must be investigated.

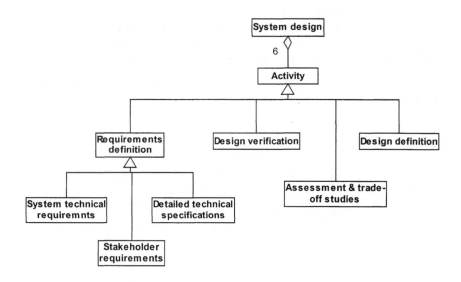

Figure 6.3 *Focus on system design*

The next step is to focus on one of the five processes, in this case 'Systems design', which is shown in Figure 6.3. Note how the original relationship from the high-level structural model in Figure 6.1 holds true here, in that 'System design' (a type of process) is made up of four (more than one) 'Activity'. Now there are four types of 'Activity': 'Requirements definition', 'Design Verification', 'Assessment and trade-off studies' and 'Design definition'. It can also be seen that 'Requirements definition' has types 'Stakeholder requirements', 'System technical requirements' and 'Detailed technical requirements'.

This modelling can also be carried out on the other four types of 'Process' that were shown in Figure 6.2. This would result in a separate diagram each for 'Acquirer supplier agreement', 'Planning', 'Control' and 'System V&V', all of which must be consistent with Figure 6.1.

At this point, we leave EIA 632 and move on to the second standard to be modelled, ISO 15504, and carry out the same level of modelling on it.

6.3.4 *ISO 15504 Software process improvement and capability determination (SPICE)*

This section looks at modelling ISO 15504, or SPICE as it is more commonly known, which will then be compared to EIA 632.

SPICE is the ISO standard that defines a set of ISO processes (that are completely compatible with ISO 9001) and relates them to a capability maturity model, similar to the CMM. The actual SPICE standard is monstrous in size, weighing in as it does at over 300 pages. This may seem fairly daunting to begin with, but SPICE itself is very well specified and is thus straightforward to model.

The first step in the model is to try to establish a high-level static view of the standard, as shown in the next diagram.

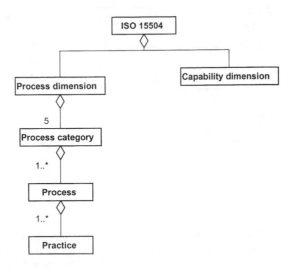

Figure 6.4   *High-level structure of ISO 15504*

It can be seen in Figure 6.4 that 'ISO 15504' is made up of a 'Process dimension' and a 'Capability dimension'. This is the highest possible level view of SPICE and the focus here will be on the 'Process dimension' rather than on the 'Capability dimension', as it is this part of SPICE that defines the process structure. The 'Capability dimension' describes the capability levels that are defined for SPICE and how they relate to practices. As this lies outside the scope of this book, the detail has been omitted from Figure 6.4.

Figure 6.4 shows that there exists a 'Process dimension' that is made up of five 'Process category'. Each 'Process category' is made up of one or more 'Process', each of which is made up of one or more 'Practice'.

The next question that needs to be asked is on which part of the model can we focus to show more detail? Note that there is an explicit multiplicity number on the association going from 'Process dimension' into 'Process category'.

Therefore, it is likely that because there is an explicit number, it may be possible to look at what these process categories are. This is shown in Figure 6.5.

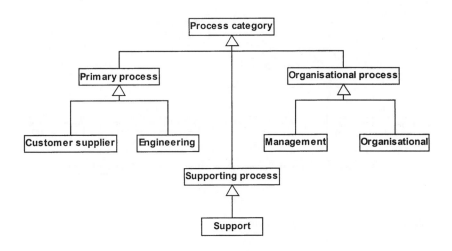

Figure 6.5 *Types of process category in ISO 15504*

Figure 6.5 shows types of 'Process category', bearing in mind that it was already stated in the previous model that there were five types of process category. From the model, there are three main types of 'Process category': 'Primary process', 'Organisational process' and 'Supporting process'. These are then further refined: 'Customer supplier' and 'Engineering' are types of 'Primary process'; 'Management' and 'Organisational process' are types of 'Organisational process'; and 'Support' is a type of 'Supporting process'. This brings the total number of actual process categories (or 'leaf' process categories) to five, as in the previous model. Notice that these middle-level process categories ('Primary process', 'Supporting process' and 'Organisational process') are conceptual groupings of processes, rather than actual categories. In UML terms, these would be described as 'abstract classes', which is to say that they are not instantiated. In UML notation, this can be represented by showing the class name in italics to indicate that it is an abstract class.

The next step is to focus in on one of these categories and examine whether the model can be taken any further.

Figure 6.6 focuses on one of the process categories defined in the previous model – in this case, 'Engineering'.

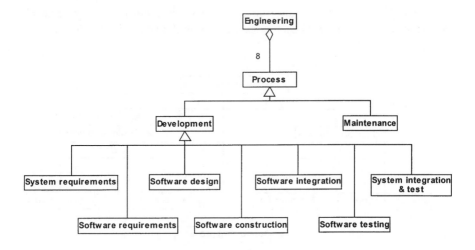

Figure 6.6 *Focus on engineering*

'Engineering' is made up of eight 'Process', which is still consistent with the original high-level structural model that stated that each 'Process category' was made up of one or more 'Process'. There are two types of 'Process': 'Development' and 'Maintenance'. 'Development' has types 'System requirements', 'Software requirements', 'Software design', 'Software construction', 'Software integration', 'Software testing' and 'System integration and test'. The process named "Development" may be thought of as being abstract and could be indicated as such on the model.

The model for ISO 15504 (SPICE) has now be modelled to approximately the same level as the EIA 632 and, at this point, the models will be compared.

6.3.5 Comparing models

The next step is to compare the two sets of models that have been created. This is simply a matter of placing the diagrams adjacent to each other and looking for similarities in terms of names and patterns.

Figure 6.7 shows both high-level models side-by-side, ready for comparison.

Figure 6.7 shows that the structure of the two standards is very similar. Both have similar-looking classes, aggregations and, indeed, even some of the terms are the same.

The most important thing to establish before the full comparison can be drawn is a common point between the two standards, from which all other relationships can be derived. The interesting thing here is that the common point occurs at the level of 'Task' (in EIA 632) and 'Practice' (in ISO 15504 (SPICE)), rather than at the common term 'Process'. This was established by looking at the definitions of the terms and also by applying common sense with regards to how they are used in the standard. Even though both standards have been modelled, it is still necessary to apply a healthy dose of common sense to make sense of, and add value to, the models.

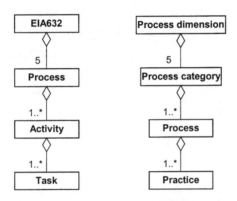

Figure 6.7   *High-level structure comparison*

This now throws up something very interesting as the next level up in the aggregation hierarchy shows that 'Activity' (in EIA 632) is equivalent to 'Process' (ISO 15504 (SPICE)) and that the term 'Process' (in EIA 632) is equivalent to 'Process category' (in ISO 15504 (SPICE)). Note that the common term 'Process' actually has a completely different meaning in both standards! This means that perhaps the most fundamental term in each standard (and one that was discussed previously as being crucial to standards), means two different things in two different standards. This has the potential to lead to a massive amount of misunderstanding and ambiguity that could prove very harmful indeed to the project.

The next step is to compare the next level of diagrams and see if any more parallels or inconsistencies can be thrown up.

It is now possible to compare the types of processes with one another, as shown in Figure 6.7, with EIA 632 being on the left and ISO 15504 (SPICE) on the right.

There is a comparison between both types of process, which can be verified by looking at the standards. In this case, the comparison is as follows:

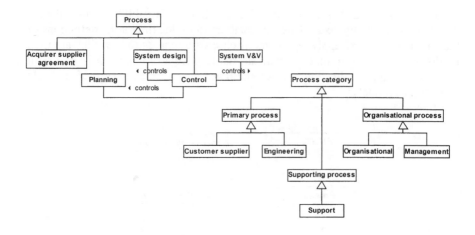

Figure 6.8   *Types of process comparison*

- 'Acquirer supplier agreement' in EIA 632 is equivalent to 'Customer Supplier' in ISO 15504 (SPICE).

- 'Planning' in EIA 632 is equivalent to 'Management' in ISO 15504 (SPICE).

- 'Control' in EIA 632 is equivalent to both 'Organisational' and 'Support' in ISO 15504 (SPICE).

- Both 'System design' and 'System V&V' in EIA 632 are equivalent to 'Engineering' in ISO 15504 (SPICE).

In this way, the common elements of both standards may be highlighted in addition to those that exist in one and not the other.

Note that the mapping is not a simple one-on-one mapping as it might first appear, but is a slightly more complex relationship. 'Control' in EIA 632 maps to two process categories in ISO 15504 (SPICE), rather than one. In the same way, 'System design' in ISO 15504 (SPICE) relates to two processes in EIA 632.

Each of the two standards may be modelled in even more detail by adding lower-level process information. For example, it is possible to model each 'Practice' that is associated with each 'Process' in ISO 15504 (SPICE) and each 'Task' for each 'Activity' in EIA 632.

6.3.5.1 Highlighting complexity

Another use for the models that are generated is to help to highlight areas of potential complexity within the standards. Although all of the models that have been generated so far have been class diagrams and, hence, static models, it is also possible to model behavioural aspects of standards using, for example, statecharts or activity diagrams. The following example shows a statechart that was generated based on the description of the acquirer supplier processes. Note that this is a statechart as each state is a normal state, as opposed to an activity or action state. This also implies that any or all of these states may contain more than one action or activity and, hence, be complex.

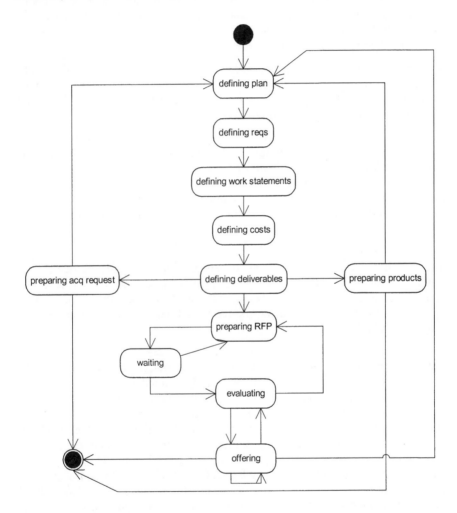

Figure 6.9 *Highlighting complexity*

The first thing that springs to mind when looking at Figure 6.9 is that the model is messy. Look at the model and try to estimate the number of possible paths through this process. What this means in systems engineering terms is that the diagram is exhibiting complexity, which may be identified in a number of ways:

- If the information that is being modelled is difficult to model, it is quite likely that complexity exists in the original information. This may be that the information is ambiguous, under-defined or over-defined. Remember that the UML is, at the end of the day, a language, and should therefore be easy to read. By the same token, if anything is difficult to model, the original source information is probably not very clear.

- If the model looks messy, complexity almost certainly exists in the model. This is a rather basic heuristic but one that tends to work. There are more formal ways to highlight complexity, such as applying metrics to the models that can actually give a quantitative result.

It may turn out that this complexity is inherent in the system (essential complexity) and, as such, nothing can be done about it. It is still important, however, to know about this complexity so that any interactions with this part of the system may be minimised. If, on the other hand, the complexity has been introduced by human error (accidental complexity), something should be able to be done to address this, such as redesigning or rewriting the original information.

6.3.5.2 Continued analysis

The models shown so far have been concerned with the high-level views of the standards. How much detail is included in the model is dependent on what level of understanding is required by the reader. It is possible to model every single aspect of a standard but there is a danger that the model would become larger than the original standard.

The analysis that has been started here may be continued by creating more statecharts that describe the behaviour of processes, but, interestingly enough, this is not always possible. Many standards are only defined at the level where they state what must be done rather than how to do it, as they tend to avoid being over-prescriptive. The fact that sometimes enough information is included in the standard to allow the behaviour or process to be modelled actually tells us something, as does the fact that sometimes a process cannot be modelled behaviourally. There is much information to be gained from modelling, even if modelling is not actually possible. When defining company procedures that may be compliant with these international standards, for example, it is often desirable to go down another level of detail and describe the behaviour of each process. Indeed, this topic will be discussed later in this chapter when defining new procedures is discussed.

It is also possible to model life cycles and life cycle models, inputs and outputs for each process and the actual structure of deliverables which will be covered later in this chapter.

6.3.6 Other standards and processes

6.3.6.1 Introduction

This section looks at a number of high-level models of other standards and processes and discusses how each one may be compared to the previous models shown in this chapter. It should be quite clear from the models that there are similarities between each of the standards as patterns begin to emerge [4]. These patterns are exactly what they say and may be identified and re-used for defining in-house procedures. These are the same as design patterns that may be re-used frequently and that will help to deliver a more robust design.

6.3.6.2 ISO 15288 Life cycle management – system life cycle processes

This standard has been developed by taking into account the experiences of the use of many other standards. Inputs to this standard include: EIA 632, IEEE 1220, ISO 15504 (SPICE) and ISO 12207. As these standards have formed an input, it is not surprising that some common patterns emerge.

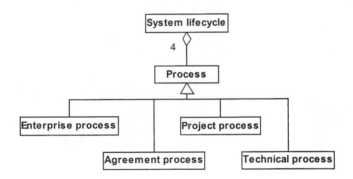

Figure 6.10 *High-level model of ISO 15288 Life cycle management – system life cycle processes*

Figure 6.10 shows that ISO 15288 Life cycle management – system life cycle processes is made up of four 'Process', with types 'Enterprise process', 'Agreement process', 'Project process' and 'Technical process'.

There is a clear mapping between this model and the two previous standards that were modelled, but note that this standard has only four main processes, whereas the previous two had five each. This means that there must be a degree of overlap between some of the types of process defined in this standard and in the others.

In addition, look at the use of the term 'Process' and try to decide exactly where this would map on to the previous two models. Would it map directly to SPICE or EIA 632, or does it have an entirely different meaning in this standard? This is exactly the type of information that we are trying to find out by modelling and it increases our understanding of the standard the more it is modelled.

6.3.6.3 IEC 61508 Functional safety of electrical/electronic/programmable electronic safety-related systems

This standard has no direct relation to any of the other standards mentioned in this section, which becomes apparent when the model is considered.

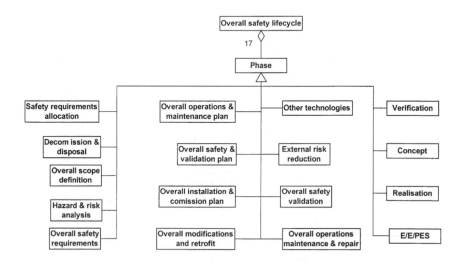

Figure 6.11 *High-level model of IEC 61508 Functional safety of electrical/ electronic/programmable electronic safety-related systems*

From Figure 6.11 it can be seen that IEC 61508 Functional safety of electrical/electronic/programmable electronic safety-related systems is made up of 17 phases.

Although at first glance this looks messy, perhaps it would look cleaner if the phases were grouped into categories. This grouping into categories is a level of organisation that exists in other standards but not in this one; this highlights that complexity exists at the highest level of the standard, whereas the lower level models turn out to be quite concise and well defined.

It is interesting to see the term 'phase' used here, which has not been mentioned so far in this chapter, but which will occur later.

6.3.6.4 IEEE 1220 Application and management of the systems engineering process

This standard has a unique pattern that stresses iteration and has been used successfully for many years, particularly in the United States. This standard has been superseded by the new ISO 15288 Life cycle management – system life cycle processes.

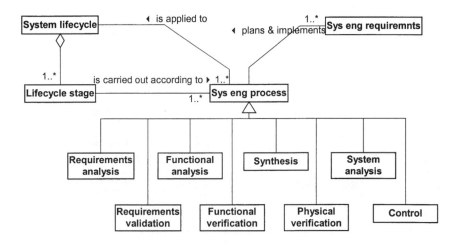

Figure 6.12 *High-level model of IEEE 1220 Application and management of the systems engineering process*

Figure 6.12 shows IEEE 1220 Application and management of the systems engineering process, which shows that a 'System lifecycle' is made up of one or more 'Life cycle stage'. Each 'Life cycle stage' is carried out according to one or more 'Sys eng process' and one or more 'Sys eng process' is applied to a 'System life cycle'. One or more 'Sys eng req' plan and implement a 'Sys eng process'. In addition, there are eight types of 'Sys eng process'.

Again, although there are more processes than in some standards, it may be due to the grouping (or lack of it) or the processes.

An interesting point here that has been raised on numerous occasions is that there is no concept of design mentioned anywhere in the model. Does this mean that the standard is not at all concerned with design, or is it an oversight, or does design go under the auspices of some other term? The reality of this is that elements of design exist within 'Functional analysis' and 'Synthesis'.

6.3.6.5   The Rational Unified Process (RUP)

The Rational Unified Process (RUP) is a natural evolution of many existing processes that is the recommended process from the originators of the UML [5, 6]. The RUP has had a large uptake in the United States and is being adopted rapidly in the rest of the world, especially Europe. One of the initial problems that some people have with the RUP is that the terminology used is 'Americanised'. This can be quite confusing for people with an ISO background, but once the mapping between terms has been established, this should not be a problem.

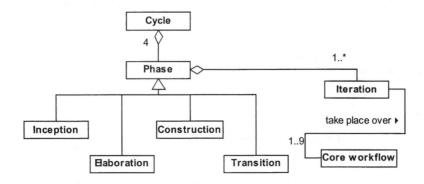

Figure 6.13 *High-level model of the Rational Unified Process (RUP)*

Figure 6.13 shows the RUP, which is recommended by the developers of the UML and which has been included here for completeness sake. It can be seen that a 'Cycle' is made up of four 'Phase', of which there are four types: 'Inception', 'Elaboration', 'Construction' and 'Transition'. Each 'Phase' is made up of one or more 'Iteration' and between one and nine 'Workflow' take place over each 'Iteration'.

If this is mapped on to the previous model, it can be seen that there is no actual term 'process', so where does the comparison start? The actual equivalent term on this model is 'Core workflow'.

6.3.6.6   EIA 731 Systems engineering capability model

This standard is intended to be a standard for systems engineering capability in the same way that ISO 15504 (SPICE) is the standard for software engineering capability determination. It is claimed to complement both IEEE 1220 and EIA 632, as well as being consistent with ISO 9001.

Like ISO 15504 (SPICE), the standard itself is huge, weighing in at several hundred pages but, also like SPICE, it is very well written and surprisingly easy to model. The high-level structure model for EIA 731 can be seen in Figure 6.14.

This standard is widely used by the defence and aerospace industries, which have used it to great effect.

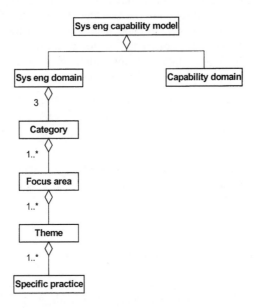

Figure 6.14 *High-level structure of EIA 731*

Figure 6.14 shows the high-level structure of EIA 731 Systems engineering capability model. It can be seen that 'Sys eng capability model' is made up of a 'Sys eng domain' and a 'Capability domain'. This pattern is very similar to the pattern in ISO 15504 and, like that standard, it is the process side of the model that we are interested in, rather than the capability side. The 'Sys eng docmain' is made up of three 'Category', each of which is made up of one or more 'Focus area'. Each 'Focus area' is made up of one or more 'Theme', each of which is made up of one or more 'Specific practice'.

6.3.7 Summary

In summary, therefore, the following points were raised concerning the use of the UML for modelling standards and processes:

- The UML is suitable for modelling standards and their associated processes. Several standards were looked at and their processes modelled, carried out at a high level.

- By modelling in this way, it is possible to highlight potential inconsistencies within a standard. Any internal inconsistencies, such as badly defined terms or inputs and outputs that do not match, would come to light when modelled visually.

- It is also possible to compare and contrast these standards once they have been modelled in order to throw up inconsistencies between standards. In the examples used here, it was quite clear from comparing the models that EIA 632 and ISO 15504 had a completely different meaning for the basic term 'process', which, if not spotted at this point, may have led to all sorts of trouble and misunderstanding later in the project.

- It is also possible to highlight areas of potential complexity within a standard. Indeed, one of the standards that was considered, EIA 632, had complexity in one of the processes, which was identified by creating a behavioural model of the process.

Now that some standards have been modelled, the next step is to use this information further to help us to define our own processes and procedures that are compatible with these standards.

6.4   Defining new processes using the UML

This section uses the information from the previous section to show how new processes can be defined that are compatible with, and fully traceable to, international standards. At the heart of this technique is, of course, modelling using the UML. The analysis models generated so far may be used in order to define a new process that may be used for in-house standards and procedures.

When more than one standard has been modelled, 'patterns' begin to emerge that are common throughout all the standards. These patterns are often the basis for defining new processes. In addition, particular patterns may be crucial for a process – such as real-time or safety-critical patterns.

6.4.1   A new procedure

Imagine the scenario now where a new procedure must be defined that is compatible with a number of international standards. This new procedure is to be used by all design teams in a systems engineering company and, therefore, it must be understood by everyone in the company and must be able to be communicated effectively for training purposes.

The two standards that have been modelled so far (EIA 632 Processes for engineering a system and ISO 15504 Software process improvement and capability determination (SPICE)) will be used as a basis for international standards compliance.

The first few steps are the same as for analysing standards, in that the high-level models are created, but then the modelling is taken further. The first step in the analysis was the high-level structure model, so this is the first model that is created for process definition.

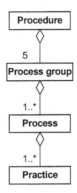

Figure 6.15 *High-level process structure model for the defined process*

Figure 6.15 shows the high-level process structure for a new procedure that is going to be defined. It can be seen that the 'Procedure' is defined as being made up of five 'Process group', each of which is made up of one or more 'Process', each of which is made up of one or more 'Practice'.

It can be seen that the terms used here are similar to the ones in the international standards that were modelled previously. This is intentional as it is intended to adopt an ISO-style terminology. Therefore, where inconsistencies do occur, the ISO term, rather than the EIA term is opted for. The exception to this is the term 'Process group', which was decided upon as it was an established term used within the organisation. However, these areas of potential misunderstanding are now known and each term can be defined explicitly at the beginning of the procedure.

Also at this point, the information in the new procedure can start to be mapped on to the analysis models for the standards, ensuring traceability and compliance with them. It should be quite clear how the high-level structure model in Figure 6.15 can be mapped directly on to both EIA 632 (Figure 6.1) and ISO 15504 (SPICE) (Figure 6.4). The next step is to define the types of 'Process group' that can be defined, which are shown in Figure 6.16.

Figure 6.16 shows the five process groups that were stated on the previous model and expands upon them.

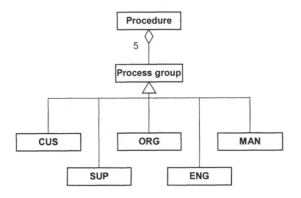

Figure 6.16 *Types of process group for the defined procedure*

There are five types of 'Process group' and these are 'CUS' (Customer/ supplier relationship), 'SUP' (support services), 'ORG' (organisational), 'ENG' (engineering) and 'MAN' (management). Again, it is worth stressing that the language here is tending towards ISO rather than EIA, which is purely a matter of preference.

It is clear to see where these process groups have come from as both of the standards that were modelled have a similar structure, or pattern.

Note also, that new terms are being defined as the models evolve, but, once more, they are fully traceable back to the original terms in the original standards.

The next step is to focus in on one 'Process group' from the previous model. The one chosen for this example is 'ENG', or engineering. The process group 'ENG' is made up of one or more processes, which holds true from the previous models. There are seven processes: 'User requirements', 'System specification', 'Design', 'Implementation', 'Integration', 'Installation' and 'Maintenance'.

Once again, it is a simple matter, at this level, to map these processes on to their corresponding elements from the two standards, simply by comparing the model in Figure 6.17 with their counterparts in Figure 6.3 and Figure 6.6.

The modelling process definition is now at the same level as the two standards that were analysed previously. However, as this is a procedure, we want to go into more detail and explicitly define which 'Practice' makes up each 'Process'.

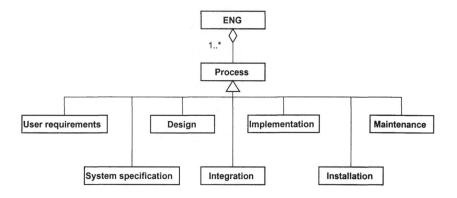

Figure 6.17   Focus on the 'Engineering' process group

The next step is to define some practices for each process, as the previous models stated that each 'Process' was made up of one or more 'Practice'. In order to make the model more manageable, these practices will be represented by operations on the class – even though they were previously shown as component classes. All that has been done here was to abstract the information to a higher level to make the models simpler.

A similar abstraction was carried out to show what documents were needed as outputs, or deliverables, for each process. This time, however, the deliverables are shown as attributes.

Figure 6.18 shows an expansion of a single process, that of 'Design'. It can be seen that seven practices have been defined and are shown on the model as the operations 'design interfaces', 'architectural design', 'detailed design', 'develop tests', 'review', 'update/issue' and 'establish traceability'. The deliverables that are associated with the 'Design' process are shown as attributes and are 'Integration test spec', 'System design', 'System test spec' and 'Design review report'. This provides us with rather a neat representation of a single process that may be shown visually by a simple class. Note that the class exhibits behaviour, or to put it another way, it has operations, and therefore it is possible to describe this behaviour using a statechart.

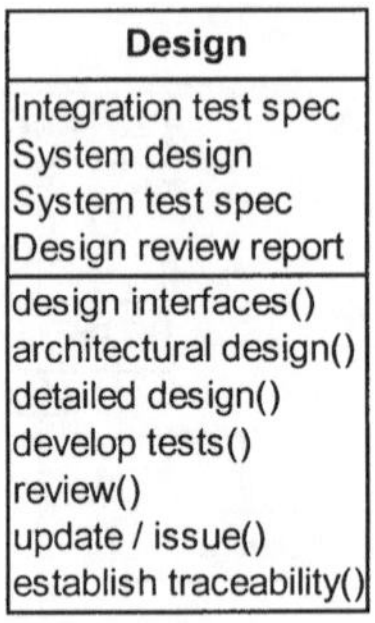

Design
Integration test spec
System design
System test spec
Design review report
design interfaces()
architectural design()
detailed design()
develop tests()
review()
update / issue()
establish traceability()

Figure 6.18 *Focus on a single process – 'Design'*

The technique used in Figure 6.18 used a single class to show a static view of a process, or what it should look like. Referring back to the UML, it had been previously stated that in order to model a system fully, any class with behaviour (operations) must have an associated statechart. The process is now described in terms of its behaviour, which shows how things are done, in what order and under what conditions, by creating a statechart for it.

Figure 6.19 shows that the 'Design' process begins by a 'start design' event, which passes the deliverable 'SS', which refers to the system specification. The first state is called 'designing high level' where three of the operations, or practices, are executed in sequence: 'architectural design', 'design interface' and 'establish traceability'. After this, a 'reviewing' state is entered where 'review' is executed and can continue in one of four ways:

- '[lo-level design incomplete]', which leads to 'designing low level' and executes 'detailed design' and 'establish traceability'. After this, control returns to the 'reviewing' state.

- '[tests incomplete]', which leads to 'developing tests' where 'develop tests' is executed. After this, control returns to the 'reviewing' state.

- '[hi-level design incomplete]', which leads back to the 'designing high level' state.

- Finally, '[complete]', which leads to 'communicating' where 'update/issue' is executed, which then leads to the final state and a 'start integration' send event.

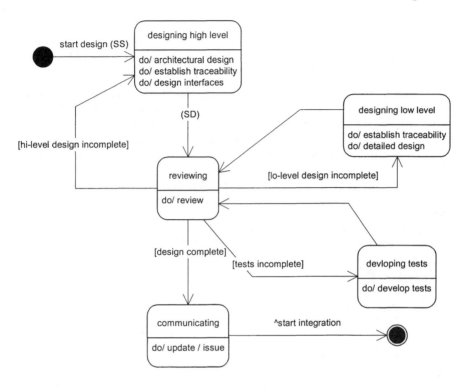

Figure 6.19 *Behavioural model for the 'Design' process*

The model shown here has complex states by placing more than one activity in some of the states. This is perfectly acceptable from a UML point of view, but some people prefer to separate them out into a 'one-activity-per-state' diagram. Although there is nothing implicitly wrong with using a single activity per state, it may be worth asking if the use of activity diagrams may be more appropriate at this point. Activity diagrams are often used to model processes or, as they are known in the RUP, workflows, so it seems appropriate to discuss them at this point. Suppose that it is possible to represent a single activity per state for the new procedure (which should be possible if it is being designed, rather than analysing what already exists), then it is possible to use an activity diagram, rather than a statechart, but are there any advantages to this approach? The answer is 'yes', as there are two previously unmentioned activity diagram elements that may be brought into use here: 'swimlanes' and 'object flow'. Swimlanes may be used to group together states and then associate them with a particular class or object. This is useful when modelling processes and workflows as it allows responsibility to be modelled on the activity diagram by relating each swimlane to a class from the stakeholder model. This also fits in with the definition of a process where responsibility must be associated with all practices.

Object flows may then be used to show the flow of deliverables around the process, by showing the object name (that will relate directly back to a deliverable) and a dependency to and from its associated states. Figure 6.20 illustrates these two new activity diagram concepts.

Figure 6.20 shows the enhanced activity diagram that represents the same process as the model shown in Figure 6.19, with a number of notable differences. Responsibility may be shown in the activity diagram by using swimlanes, which are not available for use in standard statecharts. Information flow may be shown more explicitly using the activity diagram compared to the statechart, as deliverables may be shown as objects, rather than as arguments on the statechart. There is a disadvantage, however, as it is assumed that all states are activity or action states, rather than normal states, as used in statecharts.

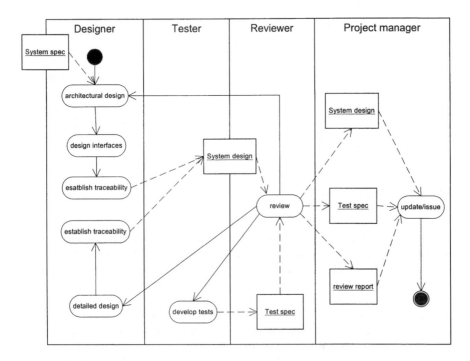

Figure 6.20 *Activity diagram with swimlanes and object flow*

Returning to the new procedure, it should be noted that this level of behavioural modelling is usually one step further than the information contained in most standards (although there are exceptions to this, the 'Acquirer supplier agreement' of EIA 632 being one of them). Standards will normally dictate what must be done in order to comply with the standard, rather than how to actually do it. This is done in order to avoid being proscriptive. With a procedure, however, it is normally desirable to be more proscriptive and to describe 'how' things must be done. If you relate this to basic UML concepts, then it fits with the definitions of static and behavioural modelling, where static modelling described the 'what' of a system and behavioural modelling described the 'how' of a system.

This level of behavioural modelling is also very useful from a project management point of view, as at any point within a project it is possible to see exactly what is being done by relating current work activities to the statechart associated with a particular process.

The model may be taken one step further by modelling the structure of an individual operation, which represents a basic practice within the procedure. These are modelled using activity diagrams and an example of such a model is shown in Figure 6.21.

Figure 6.21 shows a low-level activity diagram that models the 'architectural design' practice from Figure 6.19.

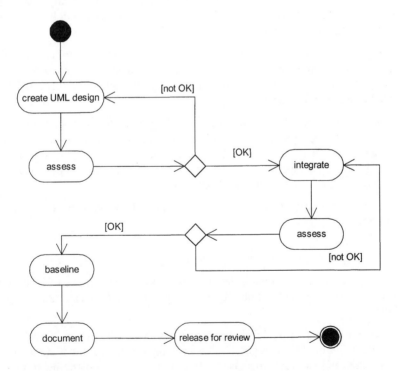

Figure 6.21 *Low-level description of the 'Architectural design' practice*

From the model, therefore, it can be seen that the first step is to 'create UML design' and then 'assess' it. If the result of the assessment is 'OK', the next step is to 'integrate'. If the result of the assessment is 'not OK', the 'create UML design' state must be revisited.

After integration, there is another 'assess' state that leads to another decision block. If the assessment is 'not OK' then 'integrate' is revisited, whereas when the assessment is 'OK' the process enters the 'baseline' state, which represents baselining the actual designs themselves. After this baselining, the designs are written up formally in a 'document' state. The final state is 'release for review', which basically makes the document available for the next state from Figure 6.19, which is the 'reviewing state'.

This model may be enhanced by adding swimlanes and object flow, in the same way that they were shown on Figure 6.20, which may add more value to the diagram.

It may be argued that modelling individual practices is going a step too far and is too much effort for the amount of work that will go into defining the models at such a low level of detail. This is, of course, true, but like most aspects of UML modelling it is entirely up to the modeller how far they go. For many situations, it may indeed be the case that this level of modelling is too detailed, but there are a number of circumstances where it may be useful:

- Imagine the situation where the design was supposed to be applied to a safety-critical application. It may be essential that the design is carried out exactly according to a set of standards or guidelines. In this situation, this low level of modelling would be useful.

- Another situation where low-level modelling is useful is where new people may be coming into a project at a late stage. In such a case, it may be a good idea to have a set of guidelines that can be made available to the newcomers to bring them up to speed on the project. Although these guidelines may not be enforceable, they are useful as a learning aid.

This is the lowest level of modelling that will be performed for process definition.

6.4.2   Completing the model

In order to complete the model, the steps that have been followed for the single process should be carried out on all the other processes. To recap, these steps are as follows:

- Define a high-level model that encapsulates the whole process structure in a single diagram. This is a good basis for comparison with other standards and processes and can form the basis of traceability for demonstrating compliance with such a standard. This model was shown in Figure 6.15 for the example of process definition.

- Categorise processes into groups that summarise the basic conceptual groupings of the processes. This can help to avoid complexity at a high level, as exists in IEC 61508, and as shown in Figure 6.11. This model was shown in Figure 6.16 for the example of process definition.

- Define practices for each process as operations on the process class. This models the basic high-level allocation of practices to individual processes. In some circumstances, such as an organisation-wide procedure, it may be desirable to cease modelling at this point. An example of this was shown in Figure 6.18 for the example of process definition.

- Define deliverables for each process as attributes on the process class. These may be derived from a class diagram that identifies deliverables per process or may simply be stated and added to the class. An example of this was also shown in Figure 6.18 for the example of process definition.

- Define a statechart or activity diagram for each process to describe the behaviour of the practices (the order in which they are carried out and the logical conditions under which the practices are executed). This will help people who need to be trained in using the procedure, or project managers who may need to know exactly what is happening (which state the system is in) at any point in time of the project. Examples of this were shown in Figure 6.19 and Figure 6.20 for the example of process definition.

- Where called for, define activity diagrams for any complex practices. This may be desirable for safety-critical applications or by way of training new staff on a project. An example of this was shown in Figure 6.21 for the example of process definition.

These steps may be repeated (or ignored) as necessary until the procedure is fully modelled. It should be quite clear by now that there is a great deal of effort involved with process definition, which is absolutely true. However, defining processes and procedures is a very complex business and it is very easy to get it wrong. By modelling such processes visually, we can have more of an understanding of the process and have more confidence in the final deliverable process or procedure.

The process models that are produced should be reviewed before any part of the actual procedure is written. Think of this as a design review for the procedure.

6.4.3 Summary

In conclusion, therefore, this section has shown how a new procedure may be modelled that makes use of the models that were generated when the standards were analysed.

This modelling was carried out in order to visualise the concepts involved in standards or procedures and has several benefits over using an English text description of a procedure:

- Traceability to the original standards may be established at the modelling stage, which is enormously useful in the event of an audit or assessment of the process. This also has the advantage that if traceability needs to be established to a new standard (as they are emerging all the time), it can be established to the process model and the procedure itself need not change.

- Early review. When creating a new procedure, it is possible to review the process at the modelling stage in order to ensure correctness of models and consistency between processes before the process is written up as a process or procedure.

- The process knowledge is stored in the process, rather than in the documented procedure itself. Imagine the written procedure as a small window through which the model may be viewed. This is useful because the procedure may be split across several documents, each one perhaps having a different audience in mind. For example, one procedure may be applicable to engineers and one to managers, one procedure may be used by subcontractors and one by in-house staff, or a different procedure may be appropriate for different types of systems, such as bespoke, niche or COTS (commercial-off-the-shelf) systems.

- The modelling does not stop here, as it is possible to model other information that may be useful in a procedure or standard such as life cycles and deliverables, both of which will be discussed in more detail later in this chapter.

This section has defined the processes and the behaviour within each, but how are theses processes executed during a real project? This is the subject of Section 6.5.

6.5 Life cycles and life cycle models

Life cycles and life cycle models are very widely misunderstood concepts. Despite the fact that they are fundamental to almost every project, life cycles and life cycle models are often confused and the terms used (incorrectly) interchangeably. There is also a great deal of misunderstanding about how processes relate to both of these concepts and, once more, people often use the term 'process' interchangeably with both life cycle and life cycle model [7].

A life cycle describes the evolution of a system from conception to its ultimate demise. The most obvious analogy to this is a human life where the life cycle starts at the actual conception, rather than birth, and continues on until death.

A life cycle is made up of a number of phases, each of which will implement one or more processes. It follows, therefore, that in order to define a life cycle, its phases must be defined. In order to define a phase, its associated process, or processes, must also be defined. Bearing in mind that a set of processes has been defined, it is possible to define a life cycle using these processes.

A life cycle model describes the configuration of a life cycle for a particular project, or to put it another way, a life cycle model defines the order in which the processes are carried out and is defined on a project-by-project basis. In UML terms, a life cycle states 'what' (a static view), while the life cycle model states 'how'.

Figure 6.22 shows these definitions using the UML. From the model, the 'Life cycle' is made up of one or more 'Phase', each of which implement one or more 'Process'. Each 'Process', as before, is made up of one or more 'Practice'.

The 'Life cycle model' shows the implementation of the 'Life cycle' and the configuration of one or more 'Phase'. This is consistent with the process definitions that were stated previously.

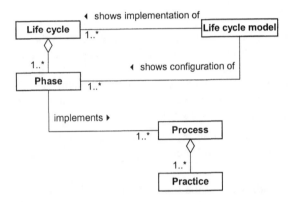

Figure 6.22 *Definition of life cycle*

Figure 6.23 shows a 'Life cycle' that is made up of one or more 'Phase'. The types of 'Phase' have been defined as 'User requirements', 'System specification', 'Planning', 'Design', 'Integration', 'Implementation', 'Installation' and 'Maintenance'.

There is no single life cycle that will apply to all projects, as it will depend on the nature of the project and its application. Phases should be chosen depending on the type of project and the processes that are available to be implemented in each phase. For example, a project that is concerned with specifying a bespoke system may include a user requirements phase, a system specification phase and an installation phase only. Another example, however, would be a project that was to be developed in-house that would include all the phases shown in Figure 6.23.

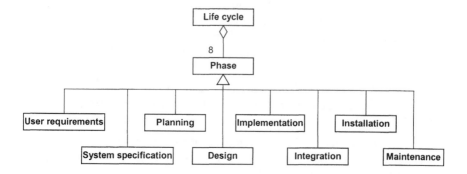

Figure 6.23 *A typical life cycle*

The phases have been named, in this case, according to the processes that they implement. This can lead to some confusion as people often assume that 'phase' and 'process' are the same thing, which is not always the case. Consider, for example, the RUP, where each phase is made up of a number of iterations, each of which may implement between one and nine core workflows. A core workflow in the RUP is equivalent to a process in ISO terms.

Note the inclusion of the phase 'Planning' here. This comes from the 'MAN' process group, which demonstrates the 'pick 'n' mix' aspect of the life cycle construction.

As this is a static view of the life cycle, the model shows what the life cycle looks like, but not how it behaves. The behaviour of the life cycle is shown by its 'Life cycle model', which may be realised in the UML using an interaction diagram.

6.5.1   Choosing a life cycle model

Choosing a life cycle model is relatively straightforward providing that the processes have been well defined. Bear in mind that a life cycle model defines the behaviour of a life cycle which is made up of phases, each of which implements one or more process. Therefore, the first step is to identify which processes will be adopted on the project. These processes should then be collected into phases and then how the phases will be executed should be determined, which is the life cycle model.

Depending on the type of project, a particular life cycle model may be selected. Many established life cycle models exist, such as 'Waterfall', 'Spiral', 'Incremental' and 'Iterative', or it may be a bespoke model. Any book on software or systems engineering will show these basic models and the way in which they are structured.

Whichever life cycle model is chosen, the phases defined in Figure 6.23 are executed. In real terms, this means executing or instantiating processes within the phase.

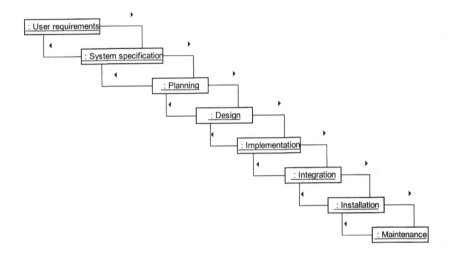

Figure 6.24 *The 'Waterfall' model*

Figure 6.24 here shows an example life cycle model known as the 'Waterfall' model. The Waterfall model allows a project to progress from one phase to the next or back to the previous phase. This sums up the life cycle model. This is perhaps the oldest of the established life cycle models and is one that has been used successfully for many years on many projects. However, the Waterfall model has also been used unsuccessfully on many projects as, like most thing in real life, it has its own limitations. To summarise, the Waterfall model is regarded as very useful when project requirements are well defined, or for small projects.

Each phase is shown as an instance of a process that has been defined previously in this chapter. It is interesting to note that one instance of a process is 'Planning', which was not defined as part of the 'ENG' process group, but is part of the 'MAN' process group.

Any type of life cycle model whatsoever may be modelled in this way using interaction diagrams (in this case, a collaboration diagram). This will dictate the permissible navigation of phases throughout a project but will not necessarily show actual project progress without introducing sequence numbers and making the model far more complex. One way to show the actual progress on a project is to show the same information, but to use the other type of interaction diagram: the sequence diagram.

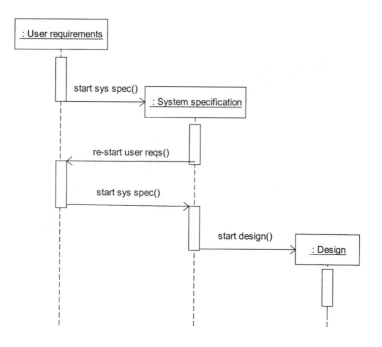

Figure 6.25 *Sequence diagram view of the Waterfall model*

Figure 6.25 shows a sequence diagram view of the Waterfall model. This model is more useful from a project monitoring point of view as it shows individual iterations of the life cycle far more clearly than the collaboration diagram model. It can be seen here that there has been an iteration in this life cycle where the progress has gone back from 'System specification' to 'User requirements', returning to 'System specification' before going on to 'Design'.

6.5.2 Summary

To summarise life cycles and life cycle models, the following points should be remembered:

- A life cycle is a static representation of a collection of phases that describe the execution of a project from its initial conception to its ultimate demise.

- A phase implements one or more process. In many life cycles, this will be a one-to-one relationship, while in others, such as iterative life cycles, this is a many-to-one relationship. It should be stressed that a process and phase are not the same thing.

- A life cycle model describes the behaviour of a life cycle over the course of a specific project.

Using the UML to represent life cycles and life cycle models is a natural progression from the process modelling that was carried out previously and is a very powerful tool for project management.

6.6 Tailoring the process

One common criticism that is levelled at any sort of process model or standard is its connection to reality. How is it possible to fully define a process yet make it flexible enough to be useable? This is often covered in standards by stating that a process must be 'tailorable', which is far easier said than done.

Depending on the type of project that a process is being used on, any or all of the processes being used may need to be specialised in some way to make them more suitable for the project at hand. This specialisation is also known as 'tailoring'. This is a useful definition of tailoring as there is a mechanism for showing special cases of classes in the UML, which is specialisation. An example of such a process is the case of a project that is concerned with safety-critical systems or real-time systems.

One practical problem that is associated with tailoring processes is that this tailoring knowledge is often lost after a project has run its course, partiularly if the tailored process was not defined well enough within the project. By using the UML it is not only possible to define how a process has been tailored, but also to retain this knowledge in the organisation by building up a process library.

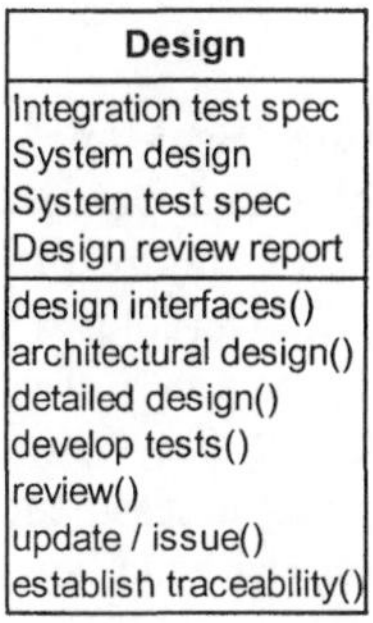

Design
Integration test spec
System design
System test spec
Design review report
design interfaces()
architectural design()
detailed design()
develop tests()
review()
update / issue()
establish traceability()

Figure 6.26 *A single process revisited*

In order to illustrate the tailoring process, an example will be used that has already been seen: – the 'Design' process from the 'ENG' process group, as shown in Figure 6.26.

The process had its deliverables defined (as attributes) and it practices defined (as operations). This is a generic engineering process that is intended to be used for generic applications. However, in some cases, this simple process may not be directly applicable to the application domain. For example, think about real-time systems or safety-critical systems where a more rigorous approach may be required. This is achieved by tailoring the process to the specific application domain.

In order to tailor a process, the class is simply specialised using the specialisation, or 'is a type of' relationship.

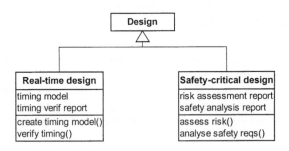

Figure 6.27 *Tailoring processes using a specialisation*

Figure 6.27 shows two specialised types or tailored processes: 'Real-time' and 'Safety-critical'.

Following the rules of UML, the 'Real-time' process will inherit all the attributes and operations (deliverables and practices) from it parent class. In addition to the deliverables and operations defined in the basic process, the 'Real-time' process also has two extra deliverables ('timing model' and 'timing verification report'~) and two new operations ('create timing model' and 'verify timing'). The same holds true for the 'Safety-critical' process, where two new deliverables have been introduced ('risk assessment report' and 'safety analysis report') and two new practices ('assess risk' and 'analyse safety reqs').

As a consequence of changing the behaviour of the process by adding new operations, a new statechart or activity diagram will have to be created.

By changing the process model by adding specialisations for each tailored process, it is possible to build up a library of tailored processes. In addition, by changing the model and keeping the specialisation in the model, the new knowledge for each tailored process is retained and may be re-used.

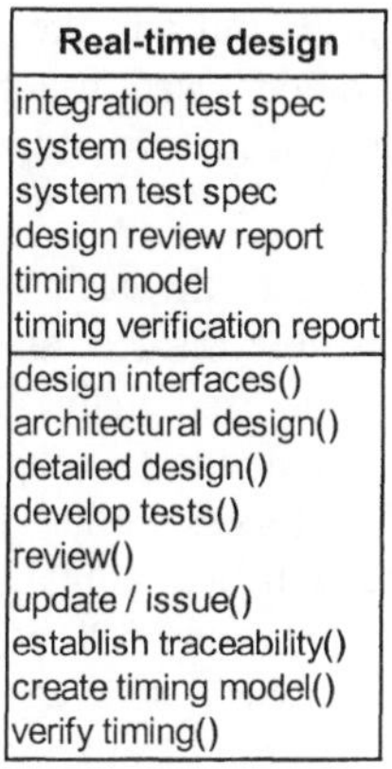

Real-time design
integration test spec system design system test spec design review report timing model timing verification report
design interfaces() architectural design() detailed design() develop tests() review() update / issue() establish traceability() create timing model() verify timing()

Figure 6.28 *The fully tailored process for 'Real-time design'*

Figure 6.28 shows the whole class for 'Real-time design', including all its inherited attributes and operations. One point to note, however, is that the number of attributes and operations is increasing. There is a danger that, as more tailored processes are defined, the processes will become unwieldy and overly complex. If this does occur, it is worth taking a step back and thinking about redefining the basic process to make it simpler. In some cases, it is even necessary to remove all attributes and operations from the basic process (which would make it non-instantiable) and to make all other processes capable of being tailored from this blank process.

6.6.1.1 Creating a life cycle

Once the tailored processes have been defined, it is possible to use them when selecting a life cycle. The procedure for this is exactly the same as before, in that a life cycle is 'assembled' from a number of pre-defined processes. However, this time there may be a choice of processes where previously there was only one option. For example, a life cycle is chosen, but this time there are three choices of process that may be implemented during the design phase: 'Design' for normal projects, 'Real--time design' for real-time projects and 'Safety-critical design' for safety-critical applications.

Any special types of project will have their application-specific knowledge retained in the 'library' of processes. This ensures that in the future, these tailored processes may be re-used and the knowledge concerning the tailoring retained in the organisation. It is often useful to attach UML notes to tailored processes to record any comments about the use of the process. This is important as practical details concerning the implementation of the process may be retained and learned from in the future. An important aspect of many standards is that lessons must be learned from each project that is carried out and they must be demonstrated. By attaching notes to processes that have been used, it is possible to demonstrate to a third party (such as an auditor) that lessons are indeed being learned from past projects.

In relation to lessons learned, consider the case where a tailored process was implemented and found to be an absolute disaster and it was decided that on no account was that particular process to be used again. One initial reaction to this is to delete it from the process library, but this is potentially hazardous for future projects. If the process is left in the process library and a note is attached to it stating never to use it, then no-one will use it again. If the process is deleted, however, there is a danger that someone may redefine it at a future date, add it back to the process library and re-use it with, once again, disastrous results. When the process is deleted, lessons are not learned from past mistakes and it is possible for history to repeat itself. By keeping the process in the library and recording negative experiences about the process, it is possible to learn from past mistakes and not make them all over again.

6.6.1.2   Summary

In summary, therefore, the following points have been illustrated:

- Tailoring a process means creating a special case of the process for a specific application or project.

- Processes are represented by classes, and thus tailoring may be represented by the specialisation UML relationship.

- Lessons learned from previous projects can be retained by saving these specialised classes in some sort of process library.

Tailoring processes covers the penultimate aspect of process modelling before the project can take place. The next step is to define deliverables, which is covered in Section 6.7.

6.7   Defining deliverables

This section looks at implementing the process by, among other things, modelling deliverables that are outputs (or work products) from a process. These may be documents, diagrams, code, physical system components, etc.

Deliverables are modelled in the same way as the rest of the standards. They may be modelled by type, process or structure.

When defining deliverables for a new process, it is possible to base them on existing analysis models from the standards, in exactly the same way as processes were previously in this chapter.

6.7.1 Deliverables by type

The first thing to look at are the types of deliverable that may exist within a project. This may be represented using a class diagram, as shown in Figure 6.29.

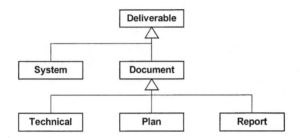

Figure 6.29 *Defining types of deliverable*

Figure 6.29 shows that the highest level of class in the diagram is 'Deliverable'. For this example, we are defining two types of 'Deliverable', which are 'System' and 'Document'. This model is by no means complete and may include concepts such as models, modelling elements, code, subsystems, etc. From the model, there are three types of 'Document' which are 'Technical' 'Plan' and 'Report'.

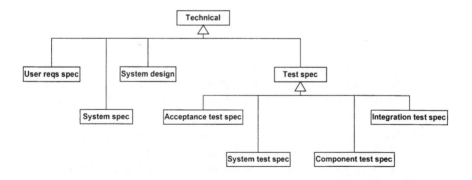

Figure 6.30 *Defining types of technical document*

The previous model may be further refined by defining types of 'Technical' document as shown in Figure 6.30. In this case, these types of technical document are 'User reqs spec', 'System spec', 'System design' and 'Test spec'; 'Test spec' is further refined by the types 'Component test spec', 'System test spec', 'Integration test spec' and 'Acceptance test spec'.

This model may be expanded to include all types of document that will exist for a project. The models here have grouped the documents by type, but it is also useful to associate these with each process, which is, effectively, looking at them from a slightly different point of view.

6.7.2   Deliverables by process

Deliverables may also be modelled by process. In fact, this has already been demonstrated once by the process descriptions, where deliverables were shown as attributes on the process class. This time, however, we shall show deliverables as classes and relate them directly to each process.

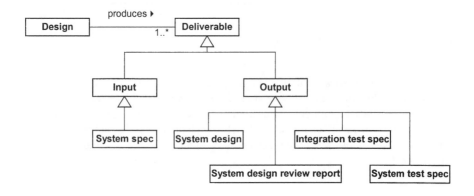

Figure 6.31 *Defining deliverables by process*

Figure 6.31 shows that 'Deliverable' has two types: 'Input' and 'Output'. There is one type of 'Input' defined as 'System spec' and four 'Output' defined as 'System design', 'System design review report', 'Integration test spec' and 'System test spec'.

Note how these deliverables also appeared on the previous model but by creating this model we are adding information to the overall model. It is now known that not only is 'System design' an output deliverable, but also that it is a technical document.

It is possible to go into even more detail by modelling the structure of deliverables and this will be discussed in Section 6.7.3.

6.7.3   Deliverables by structure

So far, the types of deliverable have been modelled along with how these relate to each of the processes that have been defined. The actual structure, or contents, of each deliverable, however, has not been discussed. It is essential that a number of guidelines are given as to the layout of each deliverable so that they can be compared and traced in a constructive fashion.

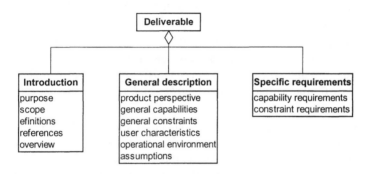

Figure 6.32 *Defining the structure of a deliverable*

Figure 6.32 shows the actual structure of a deliverable – in this case, the 'User requirement specification' – which is made up of 'Introduction', 'General description' and 'Specific requirements'. It is quite clear that these refer to sections of a document with the attributes representing subsections. This model will become the template for the user requirements specification.

Modelling deliverables in this way may seem trivial, but patterns emerge when models are compared that allow such models to be used to define new deliverables. This particular pattern for the user requirements specification deliverable will be revisited in Chapter 7 when it is used in a practical way during the user requirements phase.

6.8   Implementing the systems engineering process using a tool

So far, the theoretical aspects of defining the process have been discussed, and the next step is to implement the processes in a real-life application. There is a great deal of information contained within a process model and implementation can be very tricky. Therefore, it should be possible to implement the process using a systems engineering tool as, frankly, we can all agree that tools exist in order to make life easier.

Before the requirements of tools are discussed, it is worth considering the nature of tools, if only for a moment. A tool is exactly that, a tool. A tool will not automatically make everything right and may not even make things more efficient. It is important to look at tools with your eyes wide open and to take a really serious look at what you want out of a tool before meeting with any vendors. Many tools will be hailed as a silver bullet, or panacea, to all your problems, but this is seldom, if ever, the case. However, as long as you know what tools can do and, sometimes more importantly, what they cannot do, they can be very powerful, save much time and effort, make life easier, and make you work in a more efficient manner.

It is essential to choose a tool that will let you use your own defined processes and not just assume that the same process that the vendor uses or recommends is being used. One huge danger with many tools on the market is that they come with their own process that the tool follows. This is dangerous for a number of reasons:

- The tool is driving the process rather than the process driving the tool. Once the in-house process has been defined to reflect the way in which a particular organisation works, it is essential that this is followed, rather than some other, undefined and potentially unsuitable process that comes as part of a tool.

- By following a vendor's process, it means that the project or organisation is restricting itself to a single vendor, which is potentially expensive and, if the vendor goes bust, may pose a serious threat to the project.

- By tying in to a single vendor, it is not only the tool that will cost money, but also training and consultancy that will also be tied to a single vendor. Once you are committed to a single supplier, there is nothing to prevent the price of services from increasing dramatically.

Again, it is worth stressing that tools can be very good when used properly and for the right purpose. Anyone who has tried to hammer in a nail with anything other than a hammer (such as a pair of pliers, shoe, old house-brick or child's toy) will know that it often makes life far more difficult than walking to the tool box and getting the appropriate tool from the word go. In the same way, the tool for inserting nails is probably not appropriate for inserting windows. It is the same with systems engineering tools. A systems engineering tool that claims to perform modelling, project management and life cycle definition may end up being a jack of all trades, yet a master of none. In addition, there is nothing wrong with using more than one tool from more than one vendor to carry out different tasks. Of course, it is helpful if they work together in some way, but with the convergence of computer styles and operating systems, this is becoming increasingly practical. For a more in-depth look at tool evaluation, see Chapter 9.

6.8.1 Uses for a tool

There are many reasons for using a systems tool and the following list is a generic set of points that should apply to almost all projects:

- The tool will be used to implement the process that has been defined so far.

- The tool will be used to aid communication throughout a project by holding all project information in one place where it may be accessed by all project personnel.

- The tool will not only be used to store, but also to manage all project information in a controlled fashion.

- The tool will need traceability functions within it to ensure that all relationships between deliverables are maintained.

- The tool should be set up to hold templates for deliverables, according to the UML models, to make implementing the process as simple as possible.

One way to assess a tool's performance at a high level and to ensure that the tool will work seamlessly with a defined process, is to model the actual tool using the UML. This may seem like a strange and useless thing to do, but it helps understand the tool and the model will provide the basis for the mapping between the theoretical process and the real-life implementation of the process on a project.

Figure 6.33 is a representation of a commercial tool that is in widespread use in today's industry. Although this is by no means a complete model of the functionality of the tool, this one diagram gives an excellent overview of the tool (useful for training) and provides vital information that will be used later in this section to create the 'assumption model' and the 'implementation' model.

Figure 6.33 shows the UML diagram for a popular systems engineering tool. From the model, it can be seen that the 'Tool' is made up of one or more 'Project', each of which is made up of one or more 'Module'. There are three types of 'Module': 'Formal module', 'Link module' and 'Description module'.

Each 'Link module' is made up of one or more 'Linkset', each of which is made up of one or more 'Link'.

Each 'Formal module' is made up of one or more 'Object'. Two 'Object' are related together by one 'Link'.

This model does not show the complete functionality of the tool – in fact, it does not show any functionality of the tool! The model is, however, very useful for a number of reasons:

- It provides a very simple snapshot of the tool and shows the structure of the information in the tool.

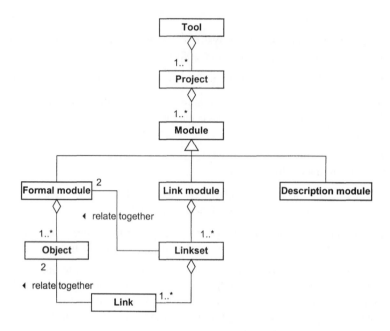

Figure 6.33 *Model of a popular systems engineering tool*

- It is useful as a training aid to give a high-level view of the tool.
- It provides the basis for successful implementation of the process. If the process is represented using the UML and the tool is represented using the UML, it is possible to map between them and create something useful.

The key to successful implementation on the tool lies in defining the 'information model' and the 'implementation model'.

6.8.2   *The information model*

The model that will actually be implemented on the tool is known as the 'implementation model'. This implementation model is based on a theoretical 'information model' that can be derived directly from the process models. The information model shows the deliverables from the process model and the relationships between them. These associations will, typically, represent traceability paths between the various deliverables, bearing in mind that traceability is a basic requirement of almost all processes and standards.

The relationships are very important here as they will determine how the different deliverables will be linked on the tool. For example, this will determine the traceability throughout the whole project, if the tool has traceability functions in it.

Figure 6.34 shows part of the information model for the process that has been defined so far. The information model shows that the 'User reqs spec' is traceable to the 'PRD' (project requirements definition) and the 'System spec' is traceable to the 'User reqs spec'. The 'Acceptance test spec' validates the 'User reqs spec' and the 'Acceptance test plan' plans the 'Acceptance test spec', while the 'Acceptance test report' records the results of the 'Acceptance test spec'.

Note the use of association names that will relate to relationships within the tool such as 'is traceable to', 'plans', 'tests' and 'records results of'.

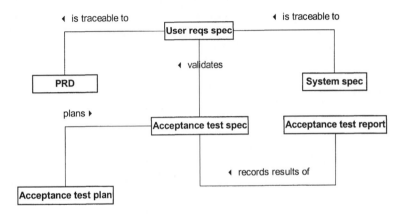

Figure 6.34 *Partial information model, based on the process that has been defined*

This model is taken from the deliverable models that were introduced in Figure 6.29 and Figure 6.31. This provides a good consistency check between the information model and the defined process.

The model here is only a small section of the entire information model. The actual information model for the 'ENG' category alone consists of over 35 deliverables and 40 associations between them. The sheer size and complexity of the information model can potentially lead to many mistakes and errors, and thus it is important that it can be translated well and effectively to the tool in order to enable the tool to manage the data.

It is essential to understand that although this model is very useful, it is still very much theoretical and does not provide all the information required to navigate a real-life project. For example, there are no explicit references to modules in the tool that has been chosen, nor any way to differentiate between different versions of the same deliverable. This information will be contained in the 'implementation model'; however, the implementation model will be almost useless with no way to map to the model of the tool. This is achieved through the 'assumption model'.

6.8.3   The assumption model

The 'assumption model' contains the key, or legend, as to how the information model will be implemented in real life using the implementation model. The assumption model provides a mapping between theory and real life: the information and implementation models.

In UML terms, the assumption model is a definition of stereotypes that will be used to make the diagram more specific to a particular application – in this case, relating the implementation of the project to the process using the UML. Stereotypes and assumption models are discussed in more detail in Chapter 8.

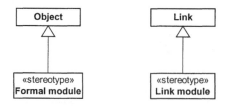

Figure 6.35 *Assumption model*

From Figure 6.35, it can be seen that two special types of class have been defined that will actually affect the UML meta-model, by defining special types (stereotypes) of basic UML entities. In this case, a stereotype of 'Object' has been defined that is called 'Formal module'. In addition, a stereotype of 'Link' has been defined, which is called 'Link module'. Notice that the two terms chosen for the stereotypes are the same terms used in the tool model, which is to say that we are now using the tool's vocabulary and introducing it into the UML using stereotypes.

This assumption model will now allow us to make sense of the 'implementation' model that will reflect real life on the project.

6.8.4   The implementation model

The implementation model is the model that shows how the theory of the process, in the form of the information model, is represented in real life using the systems engineering tool. The implementation model uses an object diagram to show how real life really exists, while still relating directly back to the defined process and using the tool's terminology.

Figure 6.36 shows the implementation model for the information model that was shown in Figure 6.35. Each object has the following features:

- An object name. This is exactly the same as the module names used on the tool. This gives a direct relation to the names of the files or modules used by the tool.

- An associated class name for each object. This is taken directly from the information model, to show the relationship to the process that has been defined using the UML.

- A stereotype name. This shows how the object is implemented on the tool – in this case, it shows whether the module is a formal module or a link module. This is taken directly from the assumption model.

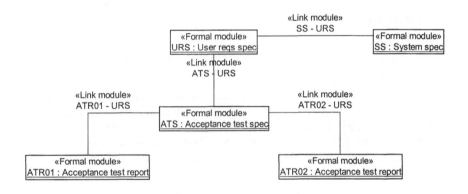

Figure 6.36 *Implementation model*

The links in Figure 6.36 show the instances of the associations from the information model in Figure 6.35 and are described in the following way:

- A link name. This is the name of the instance of the association from the information model. The actual instance name is the same as the name of the module in the tool.

- A stereotype name. This shows how the link is implemented on the tool – in this case, the link happens to be a link module.

This implementation model may be used as a navigation aid to trace between the process definition and its actual implementation on the tool.

6.8.5   Summary

This section has looked at how the process that has been defined so far in this chapter may be implemented using a basic off-the-shelf systems engineering tool.

We have now brought together the original theoretical model in the form of the information model (Figure 6.34), the model that bridges the theory and the tool in the form of the assumption model (Figure 6.35) and the actual model that is on the tool in the form of the implementation model (Figure 6.36).

6.9   Conclusions

In conclusion, therefore, this chapter has covered the following points:

- Process analysis. The UML was used to analyse and understand existing processes in the form of international standards. By modelling in this way it was possible to see any inconsistencies both within a standard and between different standards.

- Modelling new processes. The same approach was applied to creating a new process in the form of a procedure that could be traced back to the international standards upon which it was based.

- A common approach. Note the diagrams that were used for both analysis and defining a new process. Static models realised by class diagrams and behavioural models realised by statecharts, collaboration diagrams and activity diagrams. These are by no means the only diagrams that may be used because, as with all applications of the UML, any diagrams are accurate and appropriate.

- Deliverables. These were modelled by type, process and their structure to form templates.

- The process that was defined was able to be implemented using a basic systems engineering tool, which could also be modelled.

Finally, a point that cannot be stressed enough: the process must drive the tool, not the other way around. By following the approach suggested here, this is achievable.

6.10 Further discussion

1. Take the high-level process structure models from Figure 6.1 and Figure 6.4 and map them on to Figure 6.15.

2. Map between the high-level process model from Figure 6.15 and those in Figure 6.10, Figure 6.11, Figure 6.12, Figure 6.13 and Figure 6.14. Is it

possible to map between all of them? If not, why not? What could be changed to make this mapping possible?

3. Take any process from the 'ENG' process and define a number of practices in the form of operations and deliverables in the form of attributes.

4. Define the statechart or activity diagram to describe the behaviour of this new process.

5. Define deliverables for this process by their type and structure.

6. Try to model a systems engineering tool with which you have experience. If you do not use one, try to model one based on any publicly available documentation.

6.11 References

1   ERIKSSON, H. E. and PENKER, M.: 'Business modelling with UML, business patterns at work' (John Wiley and Sons Inc, 2000)

2   STEVENS, R., BROOK, P., JACKSON, K., and ARNOLD, S.: 'Systems engineering, coping with complexity' (Prentice Hall Europe, 1998)

3   EL EMAM, K., DROUIN, J. N. and MELO, W.: 'SPICE, the theory and practice of software process improvement and capability determination' (IEEE Computer Society, 1998)

4   LARMAN, G.: 'Applying UML and patterns – an introduction to object-oriented analysis and design' (Prentice Hall Inc., 1998)

5   KRUCHTEN, P.: 'The Rational Unified Process: an introduction' (Addison-Wesley Publishing, 1999)

6   JACOBSON, I., BOOCH, G. and RUMBAUGH, J.: 'The unified software development process' (Addison Wesley, 1999)

7   SCHACH, S. R.: 'Software engineering with Java' (McGraw-Hill International Editions, 1997)

Chapter 7

Modelling requirements

"what you want and what you get are two different things"
– Christine Holt

7.1 Introduction

This chapter is concerned with modelling requirements. The diagram that is most frequently used for modelling requirements in the Unified Modelling Language (UML) is the use case diagram, although, technically speaking, any diagram may be used. In this section of the book, we will be concentrating on modelling requirements using use case diagrams. See [1, 2, 3] for more discussion on use case models.

Use case diagrams, it may be argued, are perhaps one of the simplest of all diagrams, at least on the surface. However, they are perhaps the most misunderstood of all the UML diagrams. This is due in part to their inherent simplicity. Because the diagrams look so simple people often assume that very little effort is involved in generating the diagrams. This could not be further from the truth!

Requirements engineering is the discipline of Engineering that is concerned with capturing, analysing and modelling requirements. In this book we are looking purely at modelling requirements with some degree of analysis. How these requirements are arrived at in the first place is entirely up to the engineer. Indeed, many different techniques for requirements capture will be mentioned, but will not be covered in any detail, as this is beyond the scope the book.

Before the UML can be related to requirements, it is important to understand some of the basic concepts behind requirements engineering. These concepts will be introduced at a high level before any mention of the UML is made. In this way, it is possible to set a common starting point from which to work with the UML. Newcomers to the field of requirements engineering should use this section of the book as an introduction and should consult some of the books referred to at the end of the chapter for more in-depth discussion about the whys and wherefores of requirements engineering. Experienced requirements engineers should treat this section as a brief refresher, but it should be read at least once purely to determine the common starting point for the UML work.

7.2 Requirements engineering basics

7.2.1 Introduction

This section introduces some fundamentals of requirements engineering. The emphasis in this chapter is on how to use the UML to visualise requirements rather than to preach about requirements engineering itself; this section is therefore kept as brief as possible, while covering the basics. The basic concepts that are covered here will all be addressed practically in subsequent sections with respect to the UML.

Getting the user requirements right is crucial for any project for a number of reasons:

- The user requirements will drive the rest of the project and all other models and information in the project should, therefore, be traceable back to its original requirement. If you look at a typical information model, such as the one discussed in Chapter 6, this can turn out to be a large amount of documentation, all of which will be driven by the requirements.

- The user requirements are the baseline against which all acceptance tests are defined and executed. Leading on from the previous point, the success of the project will be based on the project passing the acceptance tests, which rely entirely on the project meeting the original user requirements. It follows, therefore, that the success of the project is directly based on the user requirements.

- One of the definitions of quality that ISO uses is: 'conformance to requirements'. Therefore, requirements are absolutely crucial to achieving and then demonstrating quality.

One point that will emerge from this chapter is that much of the art of analysing requirements is about organising existing requirements so that they can be understood in a simple and efficient manner.

For a more in-depth discussion of requirements engineering in general, see [4, 5, 6].

7.2.2 The requirements phase

The requirements phase is, typically, the first phase of the systems engineering life cycle model, depending on the processes that are included in the life cycle. For example, it may be that some of the customer-supplier relationship processes are implemented in a phase before requirements, but, for argument's sake, let us assume that requirements is indeed the first phase of the life cycle. According to what has been said previously in Chapter 6, the requirements phase should implement a defined process. The process that will be used in this section is taken directly from the worked examples used previously and the process is shown in the following model.

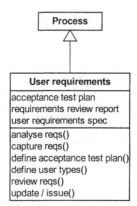

Figure 7.1   *The requirements process*

Figure 7.1 shows that there is a type of process known as 'User requirements', which has three deliverables: 'acceptance test plan', 'requirements review report' and 'user requirements spec'. The practices that must be carried out when executing requirements phase are also defined in the process as being 'analyse requirements', 'capture requirements', 'define acceptance test plan', 'define user types', 'review requirements' and 'update/issue'.

This tells us what must be done and what must be produced; it tells us nothing about how things are done, in what order and under what conditions. This is defined in the behavioural model, which is shown in the following statechart.

The statechart shown in Figure 7.2 describes the behaviour of the 'User requirements' process.

From the model, when an event called 'start project' occurs and some 'general reqs' are received (typically from the customer/supplier process) the first thing that happens is 'capturing requirements'. Three activities are carried out here: 'define user types', 'capture requirements' and 'analyse requirements'. The output of this is the 'URS' (user requirements specification), which is then used for the 'defining tests' state when the practice 'define acceptance tests' is carried out. Next, both the 'URS' and the 'ATP' (acceptance test plan) are sent to the 'reviewing' state, where the 'review reqs' practice is carried out. There are three possible outcomes from the 'reviewing' state:

- If '[URS incomplete]', the flow of control reverts back to the 'capturing requirements' state and progress continues as before.

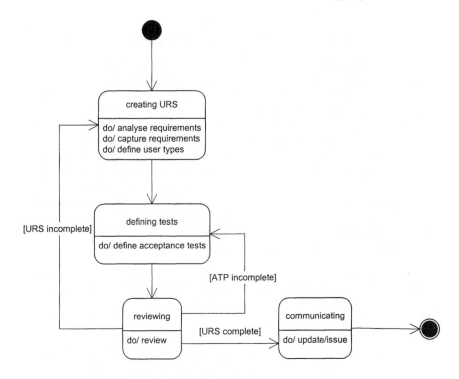

Figure 7.2 *Behaviour of requirements process*

- If '[URS complete], the 'URS', 'ATP' and 'URRR' are sent to the 'communicating' state, where the 'update/issue' activity is carried out. Finally, the baselined 'URS', 'ATP' and 'URRR' are sent, together with a 'start system spec' message, to the 'System specification' process.

- If [tests incomplete], the flow of control reverts back to the 'defining tests' state and progress continues as before.

When creating the URS, it can be seen that there are three practices that must be carried out: 'capture requirements', 'analyse requirements' and 'define user types'. The second of these practices will be discussed in the UML section as this is one of things that the UML is used for during the requirements phase. The first practice, however, is almost an entire discipline in itself and will only be mentioned briefly in the next section. The third practice, 'define user types', will be described in due course.

7.2.3 Capturing requirements

There are many ways to capture requirements, both formal and informal, and there is still much debate about which of the approaches yields the best requirements.

Rather than enter into this debate, this following list will simply give an overview of the types of technique that may be adopted when capturing requirements.

- Interviews, both formal and informal. Formal interviews follow a pre-defined format and have pre-defined questions that yield results that are comparable. This may use pre-defined forms that define a set of questions and may indeed even typical answers from which requirements may be drawn up. Informal interviews, on the other hand, start with a blank sheet of paper that will be used to record information and comments during an informal discussion. These results will then be analysed and from these, a set of requirements may be drawn up. Each of these has its own set of advantages and disadvantages. For example, formal interviews may be conducted in such a way that the questions given are quantifiable so that they may be directly compared or may be analysed statistically. On the other hand, although not so easily quantifiable, informal interviews may pick up requirements that may have been missed out by the formal questionnaire-style approach.

- Business requirements. User requirements may be derived from business requirements. A business requirement is a very high-level requirement that drives the business rather than the product, yet may drive particular aspects of the project or set up some constraints. Business requirements are discussed in more detail in due course.

- Comparable systems. Other systems with similar features may be looked at in order to capture their functionality, which may be transformed into user requirements. These other systems may be similar systems within the same organisation, legacy systems that may need to be updated, or they may even be a competitor's product.

- Maintenance feedback. The maintenance phase of a project should allow for customer feedback associated with a product. This is particularly important where the project is concerned with updating or replacing an existing system. The actual feedback may be in the form of error reports, customer wish-lists or periodic surveys of customers' opinions of a product.

- Working in the target environment. Spending some time in the target environment can be invaluable when it comes to generating requirements. It may be that users of a system simply assume some functionality that may be overlooked when it comes to stating the system requirements. A fresh viewpoint from a non-specialist may be exactly what is needed to draw out some obvious, yet crucial, system requirements.

- User groups. Many companies with a large customer base set up special user groups and conferences where users can get together and exchange viewpoints and ideas. Although expensive, these meetings can prove invaluable for obtaining customers' requirements. These meetings also demonstrate to the customer that the supplier actually cares about what customers think and has put some effort and investment into constantly improving the product that they already use.

- Formal studies. Relevant studies such as journal and conference publications can often show up 'gaps' in products or provide a good comparison with other similar products. It is important, however, to assess how much detail the studies go into as some can be superficial. It is also important to find out the driving force behind studies, as any study carried out or financed by a product vendor may lead to distorted facts and, hence, any derived requirements may be suspect.

- Prototypes. Prototypes give the customer some idea of what the final product will look like and hence can be useful to provide valuable early—life cycle feedback. Indeed, this approach is so well promoted by some that it has led to the definition of a very popular life cycle model: the rapid prototype model. The use of prototypes is particularly tempting for systems with a large software content, or those that use a computer-based interface to control or operate a system. User interface packages are very simple (and cheap to come by) and can give potential users a perfect picture of what they will be getting at the end of a project. Non-software prototypes are also very useful but can be very expensive if physical models have to be constructed.

- User modifications. These are not applicable to many systems, but some systems are designed so that they can be extended or modified by the customer. In order to relate this to a real-life example, think of a drawing or CAD package on a computer. Many of these packages allow the user to specify bespoke templates or drawing elements that may make the life of the user far simpler. Indeed, some UML tools currently on the market are basically drawing packages with pre-defined UML templates, and if getting the drawings looking good is a main requirement, this can often be the most economical solution to buying a CASE (computer aided/assisted software engineering) tool (see Chapter 9 for more information on selecting CASE tools). Whenever a modification is made, either by user or supplier, there is generally a good reason why it has been done. This reason will almost invariably lead to a new requirement for the system, which can be incorporated into subsequent releases of the system.

These are by no means a complete set of requirements capture techniques, but they give a good idea of the sheer diversity of techniques that have been adopted in the past. It may be that a completely different technique is more appropriate for your system – in which case use it!

Capturing requirements is all very well, but what in fact are requirements and what properties should they possess? The next section seeks to address some of these issues generally, before the UML is used to realise them.

7.2.4   Requirements

7.2.4.1   Overview

The previous section has introduced a number of techniques for capturing requirements, but there is more to requirements than meets the eye. For example, requirements are often over-simplified and one of the main reasons for this is because there are actually three types of requirements. It has already been stated that much of the skill involved in modelling requirement is involved with organising requirements into logical and sensible categories. These three categories are dictated by the types of requirement that are shown in Figure 7.3.

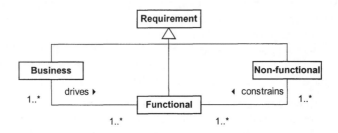

Figure 7.3   *Types of requirement*

Figure 7.3 shows that there are three types of 'Requirement': 'Business', 'Functional' and 'Non-functional'. One or more 'Business' requirement drives one or more 'Functional' requirement, and one or more 'Non-functional' requirement constrains one or more 'Functional' requirement.

'Functional' requirements are what are thought of as traditional user requirements and 'Non-functional' requirements are often referred to as 'constraints' or 'implementation requirements'. Each of these types of requirement will be discussed in more detail in the following sections.

7.2.4.2   Business requirements

The first type of requirement to be discussed is the business requirement. Business requirements, as the name implies, relate to the fundamental business of the organisation. Business requirements tend to be oriented more towards high-level business concerns, such as schedule and cost, rather than towards development itself and, as such, often fall outside the context of the system.

Business requirements and user requirements are often confused, but are not quite the same thing. Business requirements will relate to things that drive the business, such as 'keep the customer happy', 'improve product quality', etc. Because of this, business requirements are often implied, but rarely stated explicitly.

Business requirements will be aired from a point of view that ties it in to business processes. Chapter 6 introduced the subject of process modelling and showed that there were other concerns apart from engineering and, indeed, business requirements relate heavily to the 'Organisational' category, rather than 'Engineering'.

Figure 7.4 shows a set of business requirements that have been generated for a systems engineering organisation, which may help to clear up some of these points.

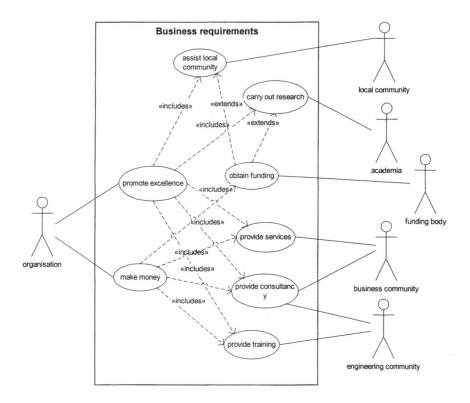

Figure 7.4 *Business requirements for a systems engineering organisation*

Figure 7.4 is actually a use case diagram that is being used to describe the business context (discussed in more detail later) for a real-life systems engineering organisation. All the use cases on this model are actually representing business requirements, rather than functional requirements. Look at the nature of these requirements and it can be seen that they exist at a very high level conceptually. For example, 'promote excellence' just happens to come directly from this organisation's mission statement, which must be a business requirement for the organisation, otherwise they are not fulfilling their mission. The second high-level business requirement is to 'make money', which must be an underlying factor associated with almost every business in the world.

The 'make money' requirement has an association with four other requirements: 'obtain funding', 'provide services', 'provide consultancy' and 'provide training'. This relationship is an 'includes' relationship, which will be discussed in more detail later in this chapter, although its meaning is self-explanatory.

The 'promote excellence' requirement also has an association with several other requirements: 'assist local community', 'carry out research', 'provide services', 'provide consultancy' and 'provide training'.

The 'obtain funding' requirement 'extends' both 'carry out research' and 'assist local community', which will be discussed in more detail later in this chapter.

The actors in this model, which are shown graphically by stick people, actually represent project stakeholders. Stakeholders represent any role, or set of roles, that has an interest in the project, and will be discussed in more detail later in this chapter.

The model that is used here will be used throughout this chapter and will drive the functional requirements for projects within the organisation. Therefore, any projects that are used as examples from this organisation must be traceable back to the company's original business requirements.

7.2.4.3   Functional requirements

Functional requirements are the typical user requirements for a system, in that they define the desired functionality of a system. User requirements should have some observable effect on a particular user of a system, otherwise they are possibly being defined at too low a level.

One of the relationships that was established in Figure 7.3 was that 'Business requirements' drive 'Functional requirements'. This means that all functional requirements should, in some way, be traceable back to the organisation's business requirements. If this is not the case, perhaps it should be questioned why this requirement is necessary.

7.2.4.4   Non-functional requirements

The third and final type of requirement that was introduced in Figure 7.3 are the 'non-functional' requirements, which are also sometimes referred to as 'implementation requirements' or 'constraints'. These 'non-functional' requirements constrain functional requirements, which means that they may limit functional requirements in some way. Examples of non-functional requirements include:

- Quality issues. A common non-functional requirement is 'comply with standard', where the word 'standard' may refer to some sort of established norm, such as an international standard or process. This is particularly relevant when read with Chapter 6, which discusses the importance of standards compliance. This may be a fundamental requirement of the system and may make all the difference between meeting acceptance tests and failing them.

- Implementation issues. It may be desirable to use a particular technique or language to implement the final system. As an example of this, imagine a system whose target environment may contain other systems and which will require a particular platform or protocol to be used.

- Life cycle issues. This may include the way in which a project is carried out. It may be that a particular project management technique should be used. An example of this is the PRINCE system, which is often cited in government projects as an essential part of the project. This may also include other constraints, such as design techniques and methodologies and even the modelling language (such as the UML!).

It is essential that these non-functional requirements are treated in the same way as both functional and business requirements, and that they exist on the requirements models.

7.2.4.5 Properties of requirements

Once requirements have been identified, it is important that they are classified in some way so that their status may be assessed at any point in the project and hey may be organised and manipulated properly. In order to classify requirements, they often have their features, or attributes, defined [7, 8, 9]. This may be represented graphically by adding, unsurprisingly, attributes to a class that represents a requirement. Figure 7.5 shows such a class with some example attributes added.

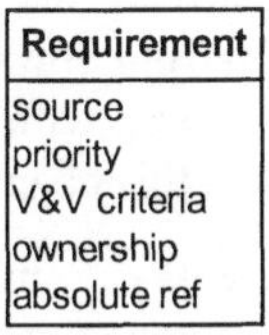

```
Requirement
------------
source
priority
V&V criteria
ownership
absolute ref
```

Figure 7.5   *Properties or a requirement shown as attributes*

Figure 7.5 shows a class that represents a requirement with several attributes that have been defined that describe the properties of a generic requirement. These requirements would be inherited by each of the three types of requirement and are particularly useful when it comes to implementing a project using a tool, which is discussed in more detail in Chapter 6.

The attributes chosen here represent typical properties that may be useful during a project; however, the list is by no means exhaustive. The attributes currently shown are:

- 'source', which represents where the requirement was originated. This is particularly useful when there are many source documents that the requirements have been drawn from, as it may reference the role of the person who asked for it.

- 'priority', which will give an indication of the order in which the requirements should be addressed. A typical selection of values for this attribute may be 'essential', 'desirable' and 'optional'. This is similar to the concept of 'enumerated types' in software, where the possible values for an attribute are pre-defined, rather than allowing a user to enter any text whatsoever. This is important, as in many cases it is simply not possible to meet all requirements in a project, and thus they are prioritised in terms of which are essential for the delivery of the system. This also takes into account the fact that many user requirements are user wishes that are no more than 'bells and whistles' and do not affect the core functionality of the system. There is often quite a large difference between what the customer wants and what the customer needs. Although both are valid requirements, it is important that the customer's needs are addressed before their wants so that a minimum working system can be delivered.

- Verification/validation criteria, where a brief idea of how compliance with the requirement may be demonstrated. This may be split into two different attributes or may be represented as one, as shown here. It is crucial that there is some clear way to establish whether a requirement may be verified (it works properly) and validated (that it has been met) as the validation will form the basis for the customer accepting the system once it has been delivered. This will be high-level information and will not be a test specification, for example.

- 'ownership', which will be related directly to the responsible stakeholder in the system. If the requirement is not owned by one of the defined stakeholders, there is something wrong with the stakeholder model. The stakeholder model is discussed later in this chapter.

- 'absolute reference', which represents a unique identifier by which the requirement may be identified and located at any point during the project. This also forms the basis for any traceability that may be established during the project to show, for example, how any part of any deliverable during any phase of the project may be traced back to its driving requirement from the user requirement specification.

Many other attributes may be defined depending on the type of project that is being undertaken. Examples of other requirement attributes include: urgency, performance and stability.

It is important that requirements are written in a concise and coherent way in order to avoid:

- Irrelevancies: it is very easy to put in too much background information with each requirement description, so that the focus is lost amidst irrelevant issues.

- Duplication: it is important that each requirement is only stated once. This is particularly important when requirements affect other requirements, such as constraints, which are ripe for duplication.

- Over-description: verbosity should be avoided wherever possible.

One of the keys to successful requirements engineering is the way in which the requirements are organised. Much of this chapter will deal with organising the models of requirements into a coherent set of usable requirements.

7.2.5   Stakeholders

7.2.5.1   Overview

This section examines another fundamental aspect of requirements modelling, which is that of stakeholders. A stakeholder represents the role, or set of roles, anyone or any thing that has a vested interest in the project, which can range from an end-user to shareholders in an organisation, to the government of a particular nation.

It is very important to identify stakeholders early on in the project as they can be used to relate to any type of requirement and can also be used to define consistent responsibilities.

7.2.5.2   Types of stakeholder

One of the practices that was identified in the requirements process was that of 'identify users'. This is often written as 'identify stakeholders' or 'identify roles' depending on which source is used for reference. The reason why these terms are often used interchangeably is that a 'user' is actually a type of stakeholder. A user will actually use the output of the project, the product, and may be thought of as having an interest in the project. Figure 7.6 shows this relationship from a UML point of view.

The relationship between stakeholders, users and the project are shown in Figure 7.1.

From the model, 'User' is a type of 'Stakeholder'. One or more 'User' uses the 'System', while one or more 'Stakeholder' have an interest in a 'Project'. In addition, the 'Project' delivers one or more 'System'.

Although it is impossible to list all possible stakeholders in an organisation, it is possible to draw up a generic model that is very useful as a starting point for a stakeholder model. Figure 7.7 shows such a generic model that will not fit all projects or organisations, but is very useful as a starting point for identifying any potential stakeholders.

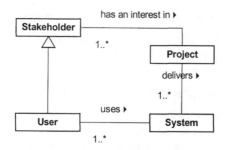

Figure 7.6   *Relationship between users and stakeholders*

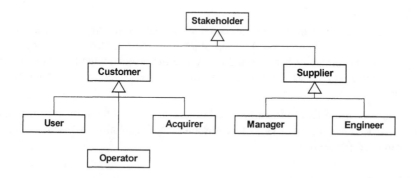

Figure 7.7   *Types of stakeholder*

It can be seen that there are two main types of 'Stakeholder': 'Customer' and 'Supplier'. These are chosen as the highest-level stakeholders as almost every project will have both a supplier and a customer. It should be pointed out that the name of the class that represents each stakeholder represents a role that is being performed, rather than the name of an individual or particular organisation. It may be that the 'Customer' and 'Supplier' stakeholders are actually within the same organisation, or may even be the same person in the case of one person who is creating a system for personal use.

The 'Customer' stakeholder has three main types: 'User', 'Operator' and 'Acquirer'. The 'User' will be the role that represents the end-user of the system – for example, in the case of a transport system such as rail or air, the 'User' role would be fulfilled by the actual passengers. The role of the 'Operator' in the same example would be taken by the staff who actually work at the stations or ports where the transport is taking place and the people who drive or fly the vehicles. The 'Acquirer' role represents whoever is responsible for paying for the project, which may be a large organisation such as an airline or rail company in the example.

The second main type of 'Stakeholder' is the 'Supplier', which has two main types: 'Manager' and 'Engineer'. Again, these are generic roles that should be used as a starting point for the real stakeholder model.

In order to take the concept of stakeholders further, let us consider an example that fits in with the business requirements that were shown in Figure 7.4, with regard to the 'provide training' business requirement. The idea here is to identify the stakeholders for this requirement by using the model shown in Figure 7.4 as a starting point for the reasoning.

Imagine a situation where a client organisation has asked the organisation whose business requirements have been defined to provide a training course at the client's premises. Therefore, a project is started that, for the sake of this example, will be referred to as the 'provide training course' project. The stakeholder model has been identified as shown in Figure 7.8.

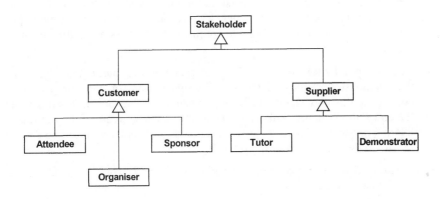

Figure 7.8 *Example stakeholder model for 'provide training requirement'*

The model in Figure 7.8 shows some stakeholders that have been identified for the 'provide training' business requirement. The two main types of 'Stakeholder' remain the same as 'Customer' and 'Supplier', because in order to deliver a training course there must be someone who wants the course, the 'Customer', and someone who can supply it, the 'Supplier'.

The 'Customer' stakeholder has been split into three types in this example:

- The 'Organiser', whose role it is to set up the logistics of the course and ensure that there are enough attendees to justify the expense of the course.

- The 'Attendee represents the people that will be attending the course.

- The 'Sponsor' represents the role that actually pays for the course.

In the same way, the 'Supplier' stakeholder has two types that differ from the original generic model. These two types are:

- The 'Tutor', who is the person or team of people who teach the course.

- The 'Demonstrator', who is the person or team of people that provide practical demonstrations that complement the tutor's teaching efforts.

It is useful to remember that the stakeholders represent the roles, rather than individuals, involved in the project. There are a number of main reasons for this:

- There is nothing to say that each role has to be a separate individual or entity, as it is possible for one person to take on more than one role. It is important that the roles are considered rather than the individual, as it may be that the system is split up with regard to the functionality attached to each role, rather than to a person.

- If an individual leaves a project and has a number of roles, it may be that more than one person can replace the initial individual. It may also occur that somebody moves position within a project, maybe up the management hierarchy, so it is absolutely crucial that the roles associated with that person are not confused with any new roles. If all associations are with an individual, rather than a role, this will become confusing to the possible detriment of the project.

- One person may take on two roles. For example, the role of the 'Tutor' and the 'Demonstrator' may be taken on by one person for one instance of the course and this may change for another instance of the course. It is important, therefore, to know the skills and specialist knowledge required by the role so that if a person who takes on both roles needs to be replaced, it can be ensured that the replacement can fulfil both roles.

This model was based on the information in Figure 7.4, but it was tailored for this particular project.

7.2.6   Summary

The information that has been introduced and discussed so far in this chapter may be summarised by the UML model shown in Figure 7.9.

In summary, therefore:

- One or more 'Requirement' are generated during 'User requirements', which is a type of 'Phase'. One or more 'Stakeholder' are identified during 'User requirements' and one or more 'Stakeholder' own one or more 'Requirement'. Each 'Requirement' has the attributes 'source', 'priority', "V&V criteria", "ownership" and "absolute ref".

- There are three types of "Requirement": "Business", "Functional" and "Non-functional". One or more "Non-functional" requirements constrain one or more "Functional requirements. One or more "Business" requirements drive one or more "Functional" requirements.

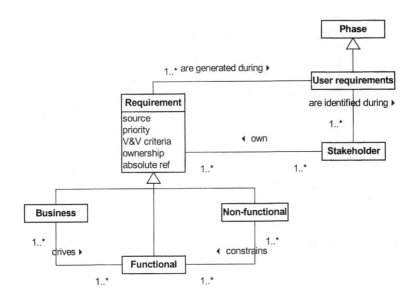

Figure 7.9 *Summary so far*

This covers the basic concepts associated with requirements engineering. These concepts will now be related to UML issues in Section 7.3.

7.3 Using use case diagrams (usefully)

Now that the basics of requirements engineering have been introduced, albeit briefly, it is useful to now take another look at the UML diagram that will be used for most of the requirements engineering models: the use case diagram. By way of a recap, the meta-model for use case diagrams is shown in Figure 7.10.

Figure 7.10 shows the partial meta-model for use case diagrams, which was first introduced in Chapter 5. It can be seen that a 'Use case diagram' is made up of one or more 'Use case', one or more 'Actor', one or more 'Association' and zero or one 'System boundary'. There are two special types of 'Association': 'Extends' and 'Includes'. A 'Use case' yields an observable result to one or more 'Actor'.

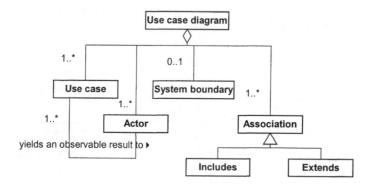

Figure 7.10 Partial meta-model for use case diagrams

7.4   Context modelling

7.4.1   Types of context

Several types of requirement have been introduced and the relationship between them defined. However, in reality, it is useful to separate these types of requirement and relate them to different types of stakeholder. This organisation of requirements results in defining the 'context' of a system.

The context of a system represents what the system is and its boundaries from a particular viewpoint. This is absolutely critical, as it is possible to have many contexts that relate to a single system, but that exist from different points of view. For example, the system context for a customer may be different from the system context of an operator. Something that is perceived as lying outside the context of one system may be included when looked at from a slightly different angle. Examples of this will be given in due course.

The context of the system shows the system boundary and any people or peripherals that communicate with the system. This may include 'who' interacts with the system and also 'what' interacts with the system. It should be noted that users are not necessarily people and may be peripherals, such as computers, hardware, databases, etc., or, indeed, another system. In terms of what has been shown so far, these entities that interact with the system are equivalent to stakeholders. This means that if all stakeholders have already been identified correctly, they can be used to help model the context of the project.

Two types of context exist: the 'business context' and the 'system context'. Each context is defined by the system boundary, which is one of the basic elements of the use case diagram. The system boundary is actually a very interesting piece of syntax for the UML as it is either massively misused or under-used. Many UML texts, tools and references ignore the system boundary altogether or simply state that it is optional. If the system boundary was optional, with no guidelines issued for when it should and should not be used, what would be the point of it in the first place? An excellent use for the system boundary element in the UML is to define a context in a system. In a nutshell, if a use case diagram has a system boundary, it is a context diagram. If it does not have a system boundary, it is simply showing the breakdown of requirements.

Each of the two types of context will now be looked at in some detail and the example used so far will be taken further and, in some cases, revisited with a new reflection on matters.

7.4.1.1 Business context

The business context shows the business requirements of an organisation or a particular project. The business requirements that were shown in Figure 7.4 are really a very high-level business context that sets the context for the whole organisation. It may be that new, additional business contexts are generated for each project, but it is absolutely crucial that these are consistent with the high-level business requirements, or context, otherwise the company will be straying from its original mission.

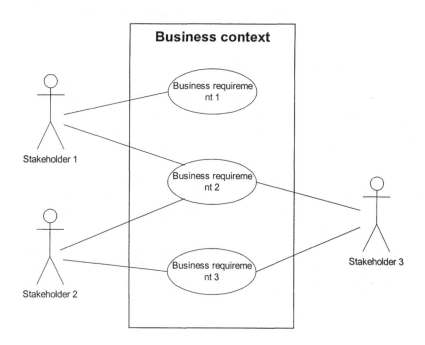

Figure 7.11 *Generic business context*

As a general rule of thumb, business requirements will exist within a business context, rather than non-functional or functional requirements. In the same vein, the roles that exist outside the context of the system are generally high-level stakeholders, rather than actual users of the system. The use case diagram shown in Figure 7.11 shows a generic business context that highlights these points.

It can be seen from the model in Figure 7.11 that the various stakeholders – 'Stakeholder 1', 'Stakeholder 2' and 'Stakeholder 3' – are associated with various business requirements: 'Business requirement 1', 'Business requirement 2' and 'Business requirement 3'.

If this is related back to the original business requirements for the systems engineering organisation in Figure 7.11, it can be seen that these rules of thumb hold true. All the actors in the diagram are stakeholders rather than users and all the use cases are business requirements.

Note the explicit use of the UML entity, the system boundary, which shows that this model is a context model rather than a straight requirements model. The system boundary is represented very simply by a rectangular box that goes around the requirements (shown as use cases) in the context and keeps out the stakeholders (shown as actors).

The second type of context is the 'system context', which is discussed in Section 7.4.1.2, before a full example is worked through, showing how to differentiate between the two types of context and offers advice on how to practically create both.

7.4.1.2   *System context*

The system context relates directly to the project at hand and is concerned with the user requirements and constraints on them, rather than with the high-level business requirements. The stakeholders that exist as actors on the use case diagram are typically the users and lower-level stakeholders, such as people who will be involved directly with the day-to-day running of the project – in other words, managers and engineers.

As a general rule of thumb, therefore, the actors on the system context will be users and low-level stakeholders and the use cases will be functional and non-functional requirements. Figure 7.12 shows a generic system context that represents these rules of thumb.

Figure 7.12 shows that the system context diagram consists of various types of functional requirement – 'Functional requirement 1' and 'Functional requirement 2 – and various types of non-functional requirement – 'Non-functional requirement 1' and 'Non-functional requirement 2'.

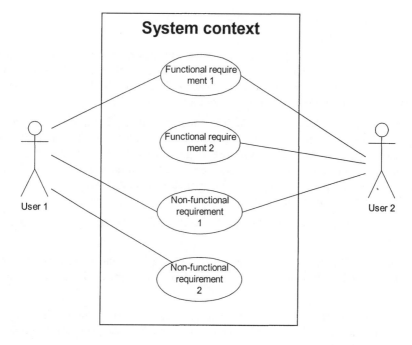

Figure 7.12 *Generic system context*

Now that the basic rules of thumb have been established, it is possible to apply these to the example that has been used previously in this chapter.

7.4.2 Practical context modelling

Consider now an example where the systems engineering organisation that has been modelled previously in Figure 7.4 wants to start up a project to run training courses. This was introduced previously and, indeed, a set of stakeholders was identified. This example will now be taken one step further and the business and system contexts will be created for the 'provide training course' project. The first question that should be asked is whether or not the project is consistent with the organisation's mission. In this case there is an obvious connection, as one of the business requirements for the organisation is to 'provide training'. If there was no clear connection to the business requirements of the organisation then one of two things must be done:

- The business requirements of the organisation must be changed to reflect the new project. If this is not done, the project does not fit in within the defined scope of the company's activities.

- If it is the case where the project is outside the scope of the company's current activities and the company is not willing to add it to its list of business requirements, the answer is quite clear: the project should not be taken on in the first place!

The first step is to ensure that the project can be proven to be consistent with the organisation's business requirements and then the next step is to identify the requirements for the project and the stakeholders. This will, in reality, be an iterative process and will take several iterations to get right.

As a first pass, let us try to identify some general requirements, regardless of type. These requirements were generated by simply asking a group of stakeholders what they perceived the requirements to be. At this stage, the requirements were not categorised, but were simply written down as a list. The list below shows a typical set of requirements that is generated when this example is used to illustrate context modelling:

- 'ensure quality', which refers to the work that is carried out by the supplier organisation.

- 'promote supplier organisation', in terms of advertising what other services the organisation can offer and hope for some repeat business or future business.

- 'improve course', which means that any lessons learned from delivering the course should be fed back into the organisation in order to constantly improve courses as time goes on.

- 'teach new skills', from the point of view of the supplier organisation.

- 'provide value', so that the clients will keep coming back for more business.

- 'organise course', which involves setting up, advertising etc.

- 'deliver course', which consists of the delivery of the course, in terms of teaching and practical demonstrations.

The next step, once the initial set of requirements has been identified, is to categorise each requirement by deciding whether it is a business requirement, a functional requirement or a non-functional requirement. The list of business requirements is:

- 'promote supplier organisation', which is clearly aimed at the business level and will relate directly back to the original business requirements of the organisation.

- 'provide value', which relates directly back to the supplier's business requirements.

The list of functional requirements is:

- 'teach new skills', which is what is expected to come out of the course.

- 'deliver course', which involves somebody turning up and teaching the course and providing course materials.

- 'organise course', which involves publicising the course, setting up the course and providing support for the course.

The list of non-functional requirements is:

- 'ensure quality', which means that the quality of the course must be maintained. In reality, this may equate to giving out course assessment forms and feeding them back into the system.

- 'improve quality', which relates to how the course may be improved constantly in order to meet one of the original business requirements of 'promote excellence'.

It may be argued that these two non-functional requirements are one and the same and, hence, would be condensed into a single requirement. It may also be argued that they are separate, distinct requirements, and that they should remain as two requirements. The problem is, which of the two is correct? As there is no simple answer to this, there is a particular rule of thumb that should be applied under such circumstances. If the answer to whether to split up or condense requirements is unclear, keep them as two. This is because experience has shown that it is far easier to condense two requirements at a later time than it is to split a single requirement in two.

Now that a number of requirements have been identified, what about the stakeholders? The list of stakeholders has already been identified in a previous section and can thus be re-used here. First of all, consider the business context and apply the rules of thumb that were discussed previously. The resulting business context is shown in Figure 7.13.

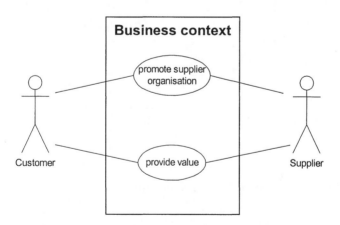

Figure 7.13 *Business context for the training courses example*

Figure 7.13 shows the business context for organising training courses. Note that the actors here are high-level stakeholders and that the use cases are business requirements. This model should be able to be traced directly back to the organisation's business requirements model. It could be argued that both use cases relate directly to both the 'provide value' and 'promote excellence' business requirements. Note also that the actors in this model are high-level stakeholders from the stakeholder model that was defined in Figure 7.8, which maintains the rule concerning high-level stakeholders and business contexts.

The next step is to create the system context, which should, hopefully, make use of the remaining requirements (both functional and non-functional) and the remaining stakeholders (user-level stakeholders). The resulting system context is shown in Figure 7.14.

Figure 7.14 shows the system context for the training courses project. This should be concerned with the nuts and bolts activities of actually running the course, rather than the high-level business requirements that appear on the business context.

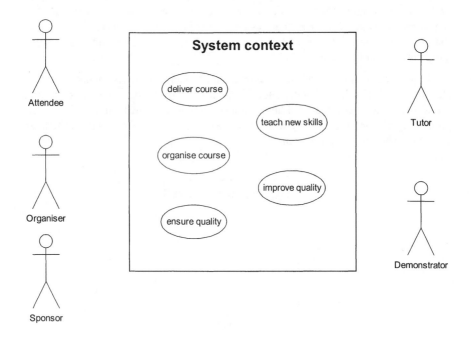

Figure 7.14 *System context for the training courses example*

The actors that have been included on this model are actually the same as the stakeholders that were identified back in Figure 7.1. By using the use case diagrams in conjunction with the original stakeholder model (realised by the class diagram) it is possible to build up some consistency between the models, even at this early point in analysing the project. One of the problems with using use case diagrams very early in the project is that there is very little to relate them to. By relating the use cases to an associated class diagram, all diagrams become more consistent and the connection to reality gets stronger. Relationships between actors and use cases have been omitted from this model as they will be discussed in more detail in due course.

So far, it has been possible to generate both the business context and the system context for a new project. The system context for the project is also traceable to the business context for the project. Note that both of these contexts are related back to the business requirements of the organisation, which is very important when it comes to establishing a business case for the new project.

7.4.3 Summary

The work covered so far in this chapter may be summarised by the following UML diagram.

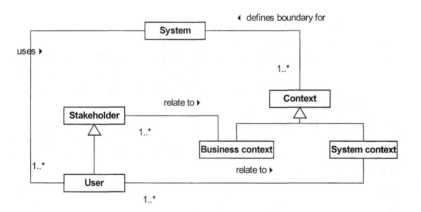

Figure 7.15 *Relationship between context and system*

In summary, Figure 7.15 can be shown to state that one or more 'Stakeholder' relate to a 'Business context' and one or more 'User' relate to a 'System context'. The 'Context' (of which there are two types) defines the boundaries of the 'System' and one or more 'Stakeholder' are identified for the 'System'.

7.5   Requirements modelling

7.5.1   Introduction

Section 7.4 was concerned with modelling the business and system contexts of the system. These contexts actually represent requirements but at a very high level. For any project, the next step would be to look at these requirements in more detail and see if they can be broken down, or decomposed, into more detailed requirements. It should be pointed out that there is nothing to say that use case diagrams have to be used for requirements modelling. Indeed, in the days before UML when people used other techniques, such as Rumbaugh's object modelling technique (OMT), class diagrams were used to model requirements. As with all aspects of the UML, you should use whatever diagrams you feel the most comfortable with. It just so happens that use case diagrams were created specifically to model requirements and, therefore, tend to be the preferred technique. In addition, this is the approach that was advocated by the old Objectory process that has since evolved into the Rational Unified Process (RUP).

7.5.2   Modelling requirements

The starting point for modelling the requirements of a system is to take the system context and use the high-level requirements contained therein. This gives an indication of what the overall functionality of the system should be. The most natural thing to do is to take each requirement and to decompose it, or break it down into smaller or lower-level requirements.

Figure 7.16 shows the systems context as generated in the previous section, except that the relationships have now been added in. Types of relationship will be discussed in some detail later in this chapter and this model will be revisited as part of the further discussion.

The first point that will be discussed, however, is that of decomposing high-level requirements into lower-level requirements.

The use case that will be taken and used for this example is that of 'organise course'. When thinking about how to decompose this requirement, it is very tempting to over-decompose. This can lead to all sorts of problems as it is quite easy to accidentally go too far and to end up taking a data flow diagram approach where the system is decomposed to such a level as to solve the problem, rather than state the requirements. This is compounded even further as use cases actually look a little like processes from data flow diagrams. Indeed, data flow diagrams were also used to define the context of a system, as are use case diagrams!

It is important, therefore, to know when to stop decomposing use cases. In order to understand when to stop, it is worth revisiting the definition of a use case, which states that a use case must have an observable effect on an actor. Therefore, consider each use case and then ask whether it actually has an effect on an associated actor. If it does not, it is probably not appropriate as a use case.

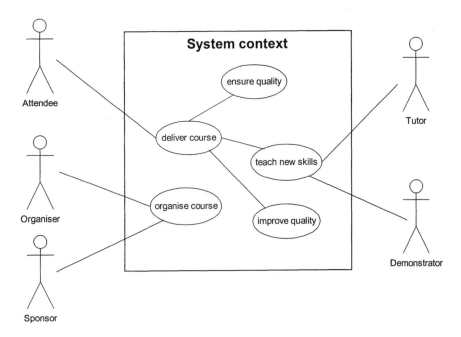

Figure 7.16 *The system context with initial relationships*

Another good way to assess whether the use case is at an appropriate level of functionality is to assess whether or not scenarios may apply directly to the use case. Remember again that a scenario is an instance of a use case and thus look to see if any scenarios are directly applicable to that use case. If no clear scenarios are apparent, it may be that the use case is too high level and needs to be decomposed somewhat. If there is only one scenario per use case, perhaps each one is too specific and the use case needs to be more generic. Scenarios will be discussed in more detail in due course.

Taking the example one step further, Figure 7.17 shows how the 'organise course' use case can be decomposed into lower-level use cases.

The model here is derived from the model in Figure 7.16, by selecting a single use case and decomposing its functionality. It can be seen now that the 'organise course' use case has been decomposed into three use cases: 'set up', 'support' and 'publicise'.

When this model is considered, imagine that one of the team also comes up with a few new use cases that are also applicable to the decomposed model, but that have slightly different relationships compared to the 'decomposed into'-style relationship that is shown in this model. The three new use cases are 'cancel course', 'organise in-house course' and 'organise external course', as shown in Figure 7.18, with their relationships to other use cases indicated by a simple associations.

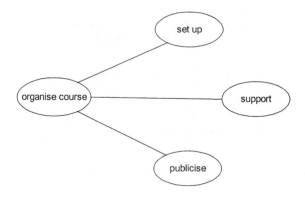

Figure 7.17 Simple decomposition of 'organise course' requirement

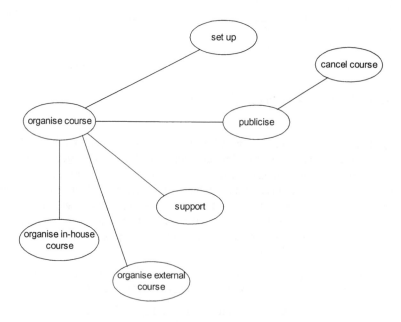

Figure 7.18 A more populated use case model

Figure 7.18 shows the updated use case model that has been populated with some new requirements. The new use cases are:

- 'organise in-house course' and 'organise external course'. These represent two different ways in which the course may be organised as sometimes the clients will come to the organisation, while on other occasions the organisation will go to the client. Although the courses will still be set up in similar ways, it is important to distinguish between the two types of course. These two requirements are directly associated with the original 'organise course' requirement.

- 'cancel course'. It is highly possible that the publicity will not be successful for the course and that too few people register, thus making the course economically unviable for the organisation. In such an event, the course must be cancelled. This new requirement is related directly to the 'publicise course' use case.

Therefore, when added to the basic decomposition-style relationship, there are two more ways in which a relationship may be defined. Clearly, these are different types of relationship and there is no way to differentiate between them, as seen in this model. This is all clearly leading somewhere and there just happen to be three types of relationship that are defined as part of the regular UML (although two of these are in-built stereotypes). These types of relationship are shown in Figure 7.19.

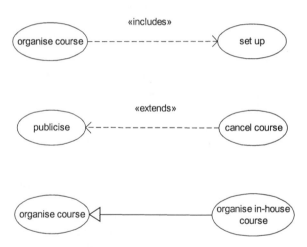

Figure 7.19 *Types of relationship*

Figure 7.19 illustrates the three basic types of relationship that may be used to add more detail to use case diagrams, which are part of the standard UML.

The '<<includes>>' association is used when a piece of functionality may be split from the main use case, for example to be used by another use case. A simple way to think about this is to think of it as an aggregation-style association. This is used to try to spot common functionality within a use case. It is highly possible that one or more of the decomposed use cases may be used by another part of the system. The direction of the arrow should make sense when the model is read aloud. This part of the model is read as 'organise course' includes 'set up'.

The '<<extends>>' association is used when the functionality of the base use case is being extended in some way. This means that sometimes the functionality of a use case may change, depending on what happens when the system is running. An example of this is when the 'cancel course' use case is implemented. Most of the time, or at least so the organisation hopes, this use case is not required. However, there may be an eventuality where it is required. This is represented by the '<<extends>>' stereotype, but it is important to get the direction of the relationship correct, as it is different from the '<<includes>>' direction. Always remember to read the diagram and the direction of the relationship makes perfect sense. In addition, note that both '<<includes>>' and '<<extends>>' are types of dependency rather than a standard association.

The final type of relationship is the generalisation relationship that is exactly the same as when used for class diagrams. In the example shown here, there are two types of 'organise course', which have been defined as 'organise in-house course' and 'organise external course'.

These new constructs may now be applied to the requirements diagram to add more value to the model, as shown in Figure 7.20.

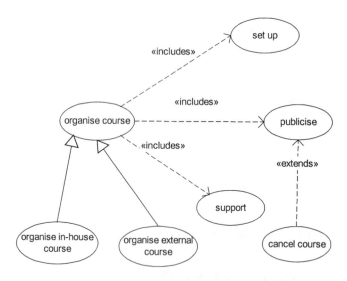

Figure 7.20 *Complete use case model for 'organise course' use case*

Figure 7.20 shows that the main use case 'organise course' has two types: 'organise in-house course' and 'organise external course'. The 'organise course' use case includes three lower-level use cases: 'set up', 'publicise' and 'support'. The 'publicise' use case may be extended by the 'cancel course' use case.

There is nothing to stop new types of relationship being defined – indeed, there are mechanisms within the UML that allow for this and these are discussed in greater detail in Chapter 8. However, let us consider a simple relationship that may be useful for requirements modelling and see how it may be applied to the example model.

It was stated previously that it is very important to distinguish between the different types of requirement that exist, whether they be 'Functional', 'Non-functional' or 'Business' requirements. However, it is unclear which of the requirements are which in some respects. We already know that anything on a business context will exclude the type 'Functional' and that any requirement in a system context will exclude 'Business' requirements, but what about 'Non-functional' requirements that may exist in both contexts? In order to help to understand this, the original definition of what exactly a 'Non-functional' requirement is must be revisited. The original definition, according to Figure 7.3, was that 'Non-functional' requirements constrain 'Functional' requirements. This will form the basis for the new type of relationship between use cases. In order to define a new UML element, the concept of a stereotype is introduced. The use of stereotypes is explained in detail in Chapter 8, but for the purposes of this example it is enough to assume that it is possible to define a new type of UML element and then, whenever it is encountered, an assumption is made about its semantic meaning. This assumed meaning is conveyed using an 'assumption model' as shown in Figure 7.21.

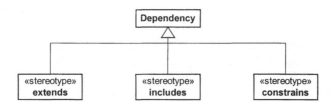

Figure 7.21 *Defining new stereotypes for use case relationships*

Figure 7.21 shows not only the existing stereotypes of relationships, but a new one called 'constrains'. Whenever the term '<<constrains>>' is encountered in a model, it is assumed that we are using a new type of UML element whose definition must be explicitly recorded somewhere, ideally with the assumption model. This new type of relationship stereotypes the 'dependency' element from the UML meta-model and is shown in Figure 7.22.

Figure 7.22 shows how the new UML element has been used to show that 'improve quality' and 'ensure quality' constrain the requirement 'deliver course'. Based on the previous definitions in this chapter, it is now quite clear that both 'improve quality' and 'ensure quality' are 'Non-functional' requirements.

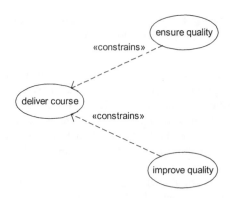

Figure 7.22 *Making 'Non-functional' requirements explicit using the 'constrains'*
stereotype

7.5.3   Describing use cases

7.5.3.1   Overview

One possible next step is to describe what happens within each use case. This may ·
be done in one of two ways: either by writing a text description or by modelling
the use case visually.

A text description is exactly what it says it is. The text description should take
the form of structured English text, much like pseudo-code descriptions from the
software world. These descriptions should describe the typical sequence of events
for the use case, including any extension points that may lead to unusual
behaviour (relating to the '<<extends>>' relationships). The text should be kept as
simple as possible with each describing a single aspect of functionality. Many
textbooks advocate the use of text descriptions, which is fine, but it may be argued
that this is going slightly against the UML ethos, which is all about visualisation.
For people who prefer visualisation, there are, of course, UML diagrams that
allow the internal behaviour of a use case to be described.

The two diagrams that are used to describe what occurs inside a use case are
the statechart diagram and the activity diagram, each of which shows a slightly
different view of the internal workings of a use case. This, however, can lead to a
certain amount of confusion when it comes to scenarios (described in more detail
later in this chapter), as scenarios show an instance of a use case and they are
modelled using either, or both, of the interaction diagrams (collaboration or
sequence).

7.5.3.2 Confusion with visual use case descriptions

In order to illustrate the confusion that can arise from describing use cases visually, consider the following meta-model description of the relationships between types of diagram.

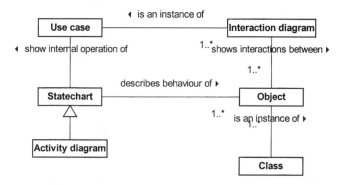

Figure 7.23 *Meta-model showing relationships between use cases and types of diagram*

It has already been stated that a scenario is an instance of a use case, and that an interaction diagram describes a scenario. Therefore, the relationship shown at the top of Figure 7.23 holds true, where one or more 'interaction diagram' describes an instance of a use case. In addition, remember that interaction diagrams were said to be used to show how societies of objects interact at a high level of abstraction, which is shown on the model as an 'Interaction diagram' shows interactions between one or more 'Object'.

Now, consider the relationship between statecharts and use cases. From the model, a 'Statechart' shows the internal operation of a 'Use case'. However, a 'Statechart' also describes the behaviour of one or more 'Object'. Bear in mind the previous point about interaction diagrams describing the interactions between objects, which describe an instance of a use case, which may be described by a statechart, and try not to get a headache!

The solution to this circular definition is to conceptualise about the slightly different views that the different diagrams represent.

- The interaction diagram describes a scenario that focuses on the interactions between objects by defining the messages that are passed between them. The key to scenarios is to show interactions between the system and anything that interacts with it. In UML terms, this will equate to the system as defined by the system boundary and its external actors.

- The statechart describes what goes on inside a use case and, hence, within the system itself.

- Activity diagrams, in one sense, can be used to describe the order of execution of activities, regardless of which object they live in.

What this leads to is the fact that a single use case can be described by a statechart to show its internal operation or by an activity diagram to show how its activities are executed alongside activities from other objects. An example of a use case (an instance) is known as a scenario, which is modelled using an interaction diagram that describes the messages passed between the system and its actors.

Maybe this is why most sources of reference for the UML advocate the use of text descriptions for use cases!

7.5.3.3 Text descriptions of use cases

The first approach to describe the internal operation of a use case is to write a simple text description of the sequence of steps that must be carried out in order for the use case to function. The example in

Figure **7.24** shows a simple text description for the use case 'publicise'.

It should be quite obvious to anyone looking at this description how the use case is intended to function. The advantage to using text descriptions is that they are very simple and, when written with care, are easy to understand. On the down side, however, a slightly more complex use case may be very difficult to represent in text terms. In addition, the formal tie in to the rest of the UML model is almost non-existent, whereas a visual approach would have stronger ties.

5.3.3.1 Publicise:

> *Receive suggested dates for course*
>
> *Send out fliers for the course*
>
> *Wait for a predetermined length of time*
>
> *When time is up, assess the response*
>
> *If the response is sufficient, send details to attendees*
>
> *If response is insufficient, cancel the course*

Figure 7.24 *Use case description using text description*

The next way to describe a use case is to model it using a statechart.

7.5.3.4 Statechart description of a use case

The second approach to describe the internal operation of a use case is to model visually using a statechart or activity diagram. The model below shows a statechart representation of the 'publicise' use case that is equivalent to the text description defined in Figure 7.24.

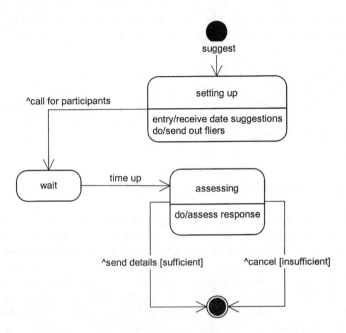

Figure 7.25 *Use case description using a statechart*

Figure 7.25 shows a visual description for the 'publicise' use case. One thing that is much clearer using the visual approach is that loops and iterations are far easy to spot. These can often be obscured, lost or misconstrued using an informal text description, but are immediately apparent when they can actually be seen. From a UML point of view, note that 'receive date suggestions' is an action whereas 'send fliers' is an activity. The keyword 'entry' applies to an action that takes zero time and is non-interruptible and so is correct for receiving a date. The keyword 'do', however, applies to an activity that indicates that the 'send fliers' will take time and may be interrupted.

7.5.3.5 *The better approach*
The immediate question that most people ask at this point is 'which is the better of the two approaches?' This can be answered with the standard UML response of 'whichever is the more suitable for the application at hand'.

Text descriptions are often preferred as they are easier to create and, it may be argued, can be very simple for anyone to understand. They also have the advantage of being in a text form which will fit directly into a user requirements document, for example. However, anything written in plain English is more prone to error and misinterpretation, simply because any spoken language is so expressive and almost all words have more than one meaning.

A visual description of a use case may be preferred for someone wanting a more formal approach to describing use cases. In addition, it is easier to reference one model to another model than it is to relate a text description to another part of the model.

Some people may not even want to describe the use cases in any way, as perhaps they do not have enough information about the requirements. One way to derive more information about requirements is to consider a scenario that interacts with the outside world, from a single aspect of functionality. Defining scenarios will be covered in more detail in Section 7.6.

7.6 Modelling scenarios

Once a set of user requirements has been generated and they have been organised and analysed using use case diagrams, it is possible to go one step further and look at 'scenarios'. A scenario shows a particular aspect of a system's operational functionality with a specific goal in mind. The definition of a scenario from the UML point of view is that it is an instance of a use case. This then gives good, strong traceability back to the original requirements as scenarios are often generated as part of the analysis during the specification phase. This is not always the case as it will depend upon the process that is being followed – indeed, it is not unheard of to have a set of scenarios defined during the requirements phase. However, as we are following the process defined in Chapter 6, it is assumed that we have now satisfied all the criteria to exit the requirements phase and have moved into the system specification phase.

Scenarios have many uses and can be a crucial part of the analysis of the user requirements and will help to identify the intended functionality of the system. This has a number of practical uses:

• It is important to identify how the user will use, and interact with, the system. Bear in mind that the actual design will take its starting point as the system specification, and thus it is important to understand how the system will be used.

• Scenarios are a good source for system and acceptance tests for a system. A scenario represents a particular aspect of a system's functionality and models the interaction between the system and the outside world, and thus they are an excellent source for defining acceptance tests.

When scenarios have been created as part of the analysis or specification phase, they will form the basis for the actual system design. Therefore, as they are traceable back to individual use cases and forward to the design, they enable a traceability path to be set up from the design right back to the requirements.

7.6.1 Scenarios in the UML

A scenario models how the system interacts with the outside world. The outside world, in the case of our system, is actually a subset of the stakeholders that have already been defined during the requirements phase and exist both on the stakeholder model as classes and also on the business and system contexts as actors. Potential paths for communication between a stakeholder and a user requirement are clearly indicated on the UML diagrams as associations that cross the system boundary.

The diagram that is used in the UML to realise scenarios is the interaction diagram, of which there are two types: collaboration and sequence. Collaboration diagrams show the interaction between objects in a system from an organisational point of view, while sequence diagrams show the interactions between objects from a logical timing point of view.

When considering scenarios, we think about actual objects in the system, rather than classes, as we are looking at a real example of operation. This gives a strong link to class diagrams as all objects must have an associated class. This is one approach to identifying classes for the analysis and design of the system.

7.6.2 Example scenarios

The first step when defining scenarios is to choose a use case from the original requirements model, which, for the purposes of this example, is shown in Figure 7.20. The use case that will be chosen for the initial example is 'publicise', which describes a requirement for the system to help the organiser publicise a training course. This is the same use case that was described previously using both a statechart and a text description.

With the use case described, it is now time to look at some possible scenarios from that use case. Two spring immediately to mind: one where publicity is successful and the course goes ahead and one where the publicity is unsuccessful and the course does not go ahead. Each will be modelled using both types of interaction diagram, which will also help to highlight the differences between the two types of diagram.

Figure 7.26 shows a sequence diagram that represents a scenario where the publicity is successful.

The first thing to notice about the model is that the main components are objects that are real-life instances of classes. Sometimes, these objects may be shown using the actor symbol if they represent stakeholders from the system context. It is usual to show the system as a single object that will allow us to identify the interactions that must occur with the outside world in order for the system to meet its original requirements.

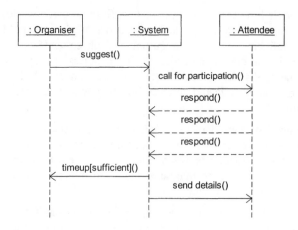

Figure 7.26 *Scenario #1 – successful publicity, sequence diagram*

Remember that the lifeline of each object, represented graphically by the vertical dotted line, shows logical time going down the page. Therefore, as time goes on, we would expect certain interactions to occur. These have already been defined to a certain extent as the use case has already been described using a statechart.

The first thing to happen is that the 'Organiser' suggests some details to the 'System'. These details may include the suggested time, date and location. The 'System' then sends out a call for participation to a number of 'Attendee'. In this scenario, a number of attendees respond to the call for participation by sending messages back to the 'System'. Eventually, the time allowed for people to respond expires and the system has to make a decision about whether to progress with the course or not. In this scenario, there is sufficient demand for the course and thus the system sends out details of the course to the attendees who have applied to attend the course. That is the end of that particular scenario.

The diagram used here was the sequence diagram (a type of interaction diagram) but there is no reason why the same information could not be shown using a collaboration diagram. To this end, Figure 7.27 shows exactly the same scenario, but modelled using a collaboration diagram rather than the sequence diagram.

The same scenario is now modelled using the other type of interaction diagram: the collaboration diagram.

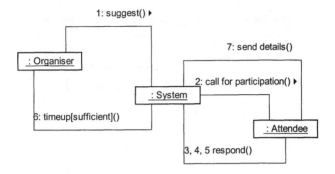

Figure 7.27 *Scenario #1 – successful publicity, collaboration diagram*

The sequence of events and the messages that are passed between two objects can still be seen, but on the collaboration diagram the emphasis is on the organisation of the objects, rather than the sequence. The actual order of events is indicated by sequence numbers that are represented by simple integers. It is possible to nest numbers by showing integers separated by points – for example, 1.1, 2.3.2 etc. It is quite clear to see that these numbers will become complicated and difficult to follow before too long.

Now consider a different scenario, one where the publicity is unsuccessful due to an insufficient number of responses. Although not frequent, this is certainly a scenario that must be taken into account when thinking about the system.

Figure 7.28 shows the second scenario. Note how the sequence of messages begins in exactly the same way as before, right up to the point when the call for participation is broadcast. The differences occur because there are not many responses this time – in fact, there is only a single response in this scenario. The same 'timeup' event occurs, but in this scenario the condition has changed. Rather than there being a 'sufficient' number of responses, there is an 'insufficient' number. The consequence of this is that the course must be cancelled, hence the 'cancel' message is sent out to any 'Attendee' who may have responded to the original call for participation.

The second scenario is shown again in Figure 7.29, but with a collaboration diagram rather than a sequence diagram. As with the sequence diagrams for both scenarios, there are similar patterns between the two collaboration diagrams. Spotting these patterns can be quite beneficial to the design process as they help to identify common areas of functionality of the system.

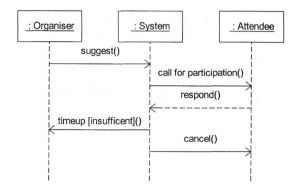

Figure 7.28 *Scenario #2 – unsuccessful publicity, sequence diagram*

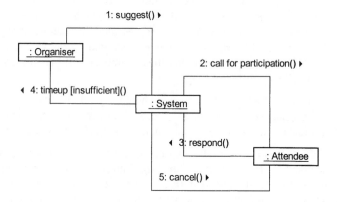

Figure 7.29 *Scenario #2 – unsuccessful publicity, collaboration diagram*

The scenarios considered so far have been deliberately simple. It is often a good idea to start off with simple scenarios that are well defined before complex scenarios are considered.

Figure 7.30 shows a new scenario that exhibits far more complexity than the previous two scenarios. This scenario relates to the 'set up' use case and describes the normal sequence of operation. Note that this sequence diagram contains a message that is passed from, and then back to, a single object. Although these are not strictly necessary, they sometimes help the person reading the model to understand more about what is going on. The message in question here is the 'print notes' message. Although this message does not describe an interaction between objects, it was deemed as being useful to include it to make explicit the fact that there is part of the scenario where printing takes place.

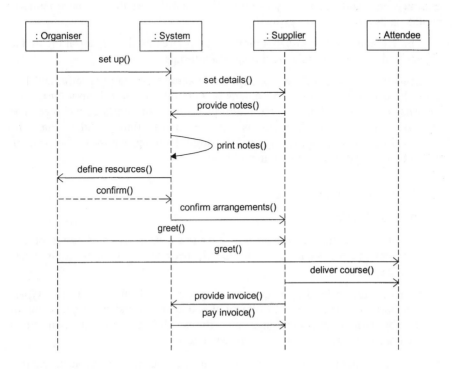

Figure 7.30 *A more complex scenario*

7.6.3 Collaboration vs. sequence diagrams

The two types of interaction diagram have been discussed and example scenarios shown. The question that is asked most frequently at this point is 'what is the point of having two diagrams that show the same information?'. This is a legitimate question and one that requires some discussion in order to fully appreciate the subtle difference between the two types of diagram.

As a starting point, it is worth thinking back to the two meta-models that were introduced in Chapter 5. The obvious difference between them is that the sequence diagram has an additional two classes as part of the 'Object' class: 'Lifeline' and 'Focus of control'. It is the lifeline that represents the flow of logical time that is missing from the collaboration diagram that emphasises the iterations of interactions between objects in the system.

Scenarios that are realised by collaboration diagrams are more generalised than sequence diagrams, particularly when there are many iterative interactions. These may be shown as a single interaction on a collaboration diagram, but may be many interactions on an equivalent sequence diagram. Iteration may be indicated on a collaboration by using a '*' symbol before the sequence number, but the paths are not shown explicitly.

The structure of the collaboration diagram stresses the organisation of the objects, rather than their relative timing relationships.

Scenarios that are realised by sequence diagrams are more specific and less generalised than in collaboration diagrams. Iterations are shown very clearly on sequence diagrams and may show return paths and more types of message. The emphasis in sequence diagrams is on the logical timing rather than the organisational structure of the system. By logical timing, we mean the order in which things happen, rather than in real time.

7.6.4   Wrapping up scenarios

In summary, therefore:

- Scenarios are defined as instances of use cases and, as use cases represent user requirements, the scenarios relate directly back to the original requirements for the system.

- Scenarios are realised by interaction diagrams, of which there are two types: collaboration and sequence. Collaboration diagrams stress the organisation of the objects in a system and their interactions, while sequence diagrams stress the logical timing relationships between objects.

- Several scenarios may be modelled for each use case. In fact, enough scenarios would be modelled in order to show a good representation of the overall system operation.

One question that is often asked about scenarios is 'how many should be produced?'. Like most of the questions that people tend to ask most often, there is a simple answer, but it is one that most people do not seem too happy with. The simple answer is 'just enough', which is very difficult to quantify. It is important to consider enough scenarios so that you have covered all typical operations of the system and several atypical operations. A good analogy here is to think about testing a system. It is impossible to fully test any complex system, but this does not mean that there is no point in testing it. In fact, it is important to test the system to establish a level of confidence on the final system. Indeed, this analogy is appropriate in another way, as scenarios form a very good basis for acceptance and system testing for just this reason.

7.7 Documenting requirements

7.7.1 Overview

This section looks at the practical issue of documenting requirements. The way in which requirements are documented will depend upon the process that is being adopted by the organisation.

Some processes do not rely on documentation as such – for example, the RUP. In the RUP, each of the models that are created and in many cases some of the components that make up the models, are stored as artefacts. These artefacts are roughly equivalent to the deliverables in a more traditional approach and form the core of the project repository. In such an approach, each use case is classed as an artefact and the models themselves are also treated as artefacts. Therefore, when it comes to documenting such use cases or models, all information is stored in the model, which can then be automatically drawn out into a report as and when necessary. This approach relies heavily on tools that are compatible with the approach and that are intended to make life simple and such documentation a matter of pressing a single key, rather than assembling documents by hand. There are advantages and disadvantages with such an approach, but these are beyond the scope of this book.

For the sake of consistency, the example that will be used here will follow on from the ISO-type process that has been defined previously.

The actual structure of the requirements documentation that will be used for this example will be based on a model of the user requirements specification (URS) that was derived from existing standards and best practices using the approach introduced in Chapter 6. Let us suppose that, in order to define the contents of a URS, a source of information was looked for. In this case, the content of the URS will be based on the one defined by the European Space Agency [10]. This was modelled according to the guidelines laid out in Chapter 6 and resulted in the model shown in Figure 7.31.

Figure 7.31 shows the structure for the URS, as derived from the process modelling exercise that has been discussed previously. The model shows the various sections of the document as classes, and the various sub-headings are shown as attributes on each class. Of course, depending on how many levels of nested heading exist in the document, there would be an associated hierarchy of classes in the document model.

It can be seen that the 'User requirements specification' is made up of '1. Introduction', '2. General' and '3. Specific requirements'. '1. Introduction' has the subsections '1.1 Purpose', '1.2 Scope', '1.3 Definitions', '1.4 References' and '1.5 Overview'. '2 General' has six subsections and '3. Specific requirements' has two subsections. In addition, '3. Specific requirements' is made up of one or more 'Requirement'. These requirements are described according to the model defined in Figure 7.5, which is actually included in this model.

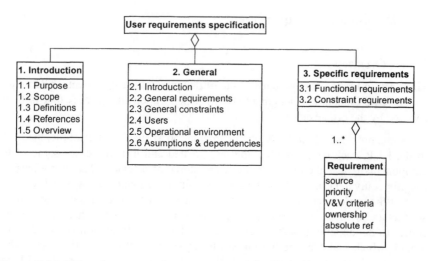

Figure 7.31 *Typical contents of a user requirements document*

This document may then be populated by making use of all the models that have been created so far, such as the contexts, stakeholder models, requirements models and scenarios.

7.7.2   Populating the document

Each part of the document that has been identified in Figure 7.31 will now be discussed and appropriate diagrams will be suggested for each section.

7.7.2.1   Section 1 – introduction

The introductory section is mainly concerned with setting the scene for the project and perhaps justifying why it is being carried out:

- The business context and parts of the system context may be used to illustrate the first two subsections of '1.1 purpose' and '1.2 scope'.

- Definitions in '1.3 definition' may be defined according to the terms used as UML elements.

- The information in '1.4 references' must be generated manually.

- Finally, '1.5 overview' may again make use of the parts of the business context and system context information.

As can be seen, most of the information for the first part of the document can be derived directly from the UML models that have been created so far.

7.7.2.2   Section 2 – general

The second section of the document may be populated as follows:

- Section '2.2 general requirements' and section '2.3 general constraints' may be obtained from high-level use cases. This will include the use case models for the high-level requirements and the business context use case model. This is because, as stated previously, many user requirements and constraints will be derived from business requirements.

- Section '2.4 users' may be defined directly from the user models and stakeholder models, which were represented previously using class diagrams.

- 'Section 2.5 operational environment' may be described by the information in the system context model, which was realised by using a use case diagram.

Again, the UML diagrams can be used to populate most of this section of the document.

7.7.2.3   Section 3 – requirements

The third section of the document may be populated as follows:

- Section '3.1 functional requirements' and section '3.2 constraint requirements' will form the bulk of the document. This information may be extracted from low-level requirement use case diagrams, requirements descriptions that were realised using text descriptions and statecharts, and, finally, scenarios. Scenarios were realised by interaction diagrams: sequence diagrams and collaboration diagrams.

The third section can be populated almost entirely using the UML models. This is not surprising, as the idea behind this chapter was to model requirements and this section is concerned purely with requirements.

7.7.3   Finishing the document

In order to finish off the document, the following points should be borne in mind:

- Include the actual models, where appropriate in the documents itself. This is not always possible but can prove to be very useful. As an interesting aside, non-UML experts will be able to understand most simple UML models with a minimum of explanation, providing that the readers do not realise that they are UML diagrams. This may sound strange, but many people will mentally 'switch off' if they think that the diagrams are in a language that they do not understand and will make no attempt to understand them.

- Add text to help the reader understand the diagrams. By basing the text directly on the diagrams, there is less room for ambiguity through misinterpretation and it also ensures that the terms used in the document and the models are consistent. It is important to remember that the knowledge concerning the project exists in the models, rather than the documents. Think of the documents as a window into the model, not the other way around.

Always remember to change the model first, rather than the document, to retain consistent project knowledge.

- Structure the actual requirement descriptions according to the structure of the requirements that was shown in the original requirement model, showing attributes associated with each requirement. Again, this leaves less room for ambiguity and, if the modelling has been performed well, should be a good way to structure things anyway.

- The document will then drive the rest of the project and be a formal view of the models. Any changes that are to be made to the system should be made in the model and then reflected into the user requirements document.

Above all, remember to make as much use as possible of the UML models. Every model will have been created for a reason, so bear them all in mind when writing the document. Modelling should be an essential, useful and time-saving part of the project, not just an academic exercise to produce nice pictures.

7.8   Summary and conclusions

In summary, therefore, this chapter has introduced and discussed the points covered by the following model.

Figure 7.32 shows the following points:

- Each 'Project' is made up of a number of one or more 'Phase'. One type of 'Phase' is 'User requirements'. No other phases were shown at this point as it was beyond the scope of the chapter.

- One or more 'Requirement' is generated during the 'User requirements' phase and each 'Requirement' must be one of three types: 'Business', 'Functional' or 'Non-functional'.

- One or more 'Business' requirement drives one or more 'Functional' requirement and one or more 'Non-functional' requirement constrains one or more 'Functional' requirement.

- One or more 'Stakeholder' is identified during the 'User requirements' phase and each owns one or more 'Requirement'. There were two main types of 'Stakeholder' identified: the 'Customer' and the 'Supplier'.

- There are two types of 'Context': the 'Business context' and the 'System context'. These types of 'Context' define the boundaries for the 'System', which is delivered by the 'Project'.

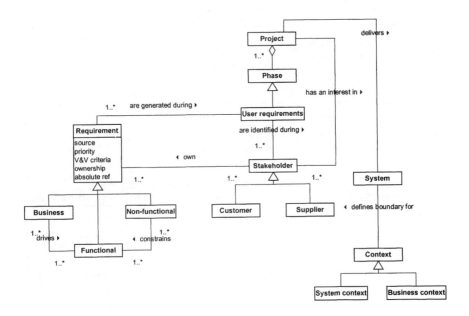

Figure 7.32 *Concepts introduced in this chapter*

This basically summarises the concepts of requirements modelling, but this chapter related all these concepts to UML modelling, which may be summarised by the following model.

The relationships between requirements engineering concepts and UML elements are summarised in Figure 7.33. From the model, it can be seen that:

- Class diagrams were used to create a stakeholder model, which was useful when trying to identify stakeholders, which is an essential part of the user requirements phase.

- Each stakeholder, once identified, could be represented as a UML actor on a use case diagram. This becomes particularly important when it comes to modelling the business and system context for a particular project.

- The context for the project, whether it is the system context or business context, is indicated by using a UML system boundary element. This element is under-used by most modellers and there is a general lack of support for system boundaries on current CASE tools.

- Use cases were used to represent the three different types of requirement. Each use case was used to represent either a single requirement or a grouping of like requirements, which could then be decomposed into further requirements.

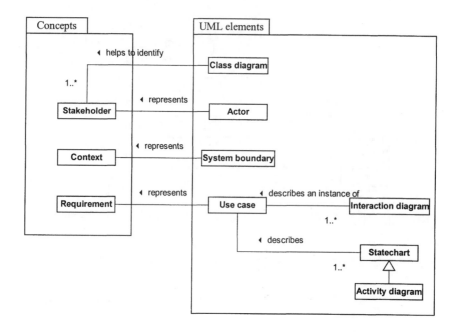

Figure 7.33 *Summary of relationships between concepts and UML elements*

- Interaction diagrams were used to model specific scenarios, in order to understand the operation of the system. Each scenario was an instance of a use case that gave a good, strong traceability path back to the original requirements model.

- Use cases and thus requirements could be further described using either a text description or a statechart or activity diagram. It is important that these are differentiated from interaction diagrams that show scenarios, or instances, of a use case.

The best way to learn these concepts in more detail, like all aspects of the UML, is to take an example and model it for yourself. Section 7.9 gives some ideas about how these ideas may be taken further.

7.9 Further discussion

1. Consider a project that you are familiar with and try to identify a stakeholder model, using a class diagram. Does the generic model shown in Figure 7.7 work for your example, or does it need to be changed considerably? In either case, why is this? Which context will each stakeholder appear in: the system context, the business context, or both?

2. Take another look at the model shown in Figure 7.16 and select a different requirement. Try to decompose this requirement into lower-level requirements so that it still makes sense. Also, bear in mind that whatever requirements are generated must be kept in line with the original business requirements that were shown in Figure 7.4.

3. Look at the requirements shown in Figure 7.20 and select a different use case from the one chosen for the example. This time, try to create some scenarios that would represent typical system operation. Use both the sequence diagrams and collaboration diagrams and compare and contrast them. Can you think of a situation where the collaboration diagram would be more useful than the sequence diagram and vice versa?

4. Choose any requirement that has been discussed in this chapter (whether functional, non-functional or business requirement) and document it as if it were part of a user requirements specification. Which part of the document would it be most applicable to?

5. Model the business requirements for another organisation (maybe your own) and see if they make sense to someone else. Choose a few existing projects or other pieces of work and see if they fit in with your business requirements model. Is the organisation's mission statement included as part of this model? Is it possible to demonstrate to a third party that the organisation is meeting its mission statement, or not?

6. Consider the system context shown in Figure 7.16. How would this be different if it were the system context from the customer's point of view, rather than the supplier's? Are there any new requirements that need to be added, or any existing ones that need to go? Are the stakeholders the same for both contexts and, if not, then why not? Does the new system context still meet the business requirements in the business context? Does it need to meet these business requirements?

7. Take one of the business requirements shown in the business context in Figure 7.4 and try to think of the type of project that may be associated with some of the other business requirements. Can you come up with a quick system context for such a model? Which stakeholders, if any, are relevant to the new system context and the business context? Are there any new stakeholders to be identified?

8. Complete the model in Figure 7.14 using the three basic and one stereotyped relationship introduced in this chapter.

7.10 References

1   JACOBSON, I.: 'Object-oriented software engineering: a use case-driven approach' (Addison-Wesley, 1995)

2   JACOBSON, I., BOOCH, G. and RUMBAUGH, J.: 'The unified software development process' (Addison Wesley, 1999)

3   SCHNEIDER, G. and WINTERS, J. P.: 'Applying use cases – a practical guide' (Addison Wesley, 1998)

4   JACKSON, M.: 'Software requirements and specifications' (Addison-Wesley Publishing, 1995)

5   DAVIES, A. M.: 'Software requirements: analysis and specification' (Prentice-Hall International Editions, 1990)

6   SCHACH, S. R.: 'Software engineering with Java' (McGraw-Hill International Editions, 1997)

7   STEVENS, R., BROOK, P., JACKSON, K., and ARNOLD, S.: 'Systems engineering, coping with complexity' (Prentice Hall Europe, 1998)

8   O'CONNOR, J., and McDERMOTT, I.: 'The art of systems thinking, essential skills for creativity and problem solving' (Thorsons, London, 1997)

9   SKIDMORE, S.: 'Introducing systems design' (Blackwell Publishing, 1996)

10 MAZZA, C., FAIRCLOUGH, J., MELTON, B., DE PABLO, D., SCHEFFER, A. and STEVENS, R.: 'Software engineering standards' (Prentice Hall, 1994)

Chapter 8

Extending the UML

"there is nothing permanent, except change"
– Heraclitus

8.1 Introduction

This chapter looks at how the UML can be extended using three basic extension mechanisms: stereotypes, constraints and tagged values. In order to illustrate these concepts an example will be considered, that of modelling websites. There are two main reasons for this: almost everyone is familiar with the world-wide web nowadays and very few people have modelled this using the UML, so it represents an interesting new application area.

Before the first example models are introduced, a rough background to the rationale behind modelling websites will be discussed. Once the scene has been set for the modelling, some examples will be introduced.

The first examples will be concerned with defining and applying stereotypes to UML elements for a particular example, in this case websites. This example is then taken further by adding constraints to the existing models in order to add new rules to the UML. The final type of extension mechanism, tagged values, shows how extra information may be added to UML models that may help with outside tasks, such as project management and version control.

The concepts discussed in this chapter have already appeared at various points throughout the book and the reader is encouraged to revisit these models after reading this chapter.

8.2 Modelling websites

8.2.1 Background

Since the advent of the world-wide web, or WWW, in 1993 the world has seen an enormous growth in the Internet. Although the Internet has been around since the

1960s it has taken the advent of the WWW and the original Mosaic browser to really get the general public interested.

The WWW, therefore, is arguably one of the fastest growing areas both in terms of technology (browsers, Java, networks, operating systems) and in terms of the growing number of users.

It is not only technology-based companies that are exploiting web-based technology, but also companies of all sorts, organisations, schools and individuals. In fact, just about everything has its own website and part of the reason for this is that websites are, on one level, very easy to build, although how useful these are is another matter entirely.

There is an argument therefore that there is no need to actually 'design' websites for a number of reasons:

- Websites are easy to write. The language of the web, the hypertext mark-up language (HTML) is very simple to use. It is based on ASCII text and requires no complicated file formats. Indeed, the number of 'instructions' is very low, which means that it is easy to learn. In fact, anyone with a text editor can create a website.

- Many tools are available. These range from full web-specific tools to word processing packages that allow documents to be stored as HTML files. By writing web pages this way the author does not even need to see, let alone type, any HTML.

- No specific knowledge is required. Adding to these reasons is that many sites have few pages and anyone who can type can create them. In many cases, the websites will only consist of basic text and images and will not be very sophisticated, which may be a reflection on the small amount of knowledge required to create them. At the other extreme, however, are websites that try to make use of every aspect of the HTML and the resultant website can be best described as difficult to manage.

There are, however, a number of arguments against these points:

- By modelling, it is possible to define templates that will ensure a similar format for all pages. This will be discussed in more detail later. These templates may also be applied retrospectively to verify the structure of a website.

- As technology advances, many web pages are not just simple HTML but have an active component such as Java or ActiveX. As websites move from being a simple HTML-based website to a full-blown web-based application that may very well interact with other parts of a system, such as databases, directories, files and other programs or applications [1], the more issues are raised concerning security. As this happens, there is a clear need for proper design models.

All this seems startlingly familiar as anyone who remembers the software crisis will appreciate. The software crisis arose when people first got their hands on computers and, just because they could type, they assumed that they were computer programmers. Brooks sums this up rather nicely with his 'sorcerer's apprentice' analogy, where a foolish cartoon mouse mistakes sorcery for hand waving, book reading and hat wearing, with disastrous results [2].

With the rationale for why websites need to be modelled firmly established, it is worth turning the focus to HTML – the language of the world-wide web.

8.2.2   Modelling HTML

This section looks at modelling the HTML language using the UML. This is a particularly interesting example to model as HTML is a simple and very well-defined language. In fact, at the highest level, HTML can be modelled with one very simple model, as shown in Figure 8.1.

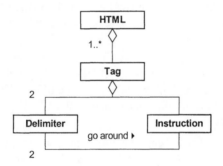

Figure 8.1   *High-level model of HTML*

Figure 8.1 shows how the HTML may be modelled using the UML at a very high level of abstraction.

It can be seen that 'HTML' is made up of one or more 'Tag'. Each 'Tag' is made up of two 'Delimiter' and one 'Instruction'. The two 'Delimiter' go around the 'Instruction'.

In a nutshell, this is the essence of HTML, which may now be expanded upon to include the whole of the HTML language consisting, as it does, of a number of different types of tag.

Figure 8.2 shows a model of a small part of the HTML language. By focusing on the 'Tag' class it is possible to model the many types of 'Tag' that exist. From the model, there are two types of 'Tag': 'Body' and 'Frame'. Of course, not all Tags are shown here due to the limited space, but two examples are shown. This

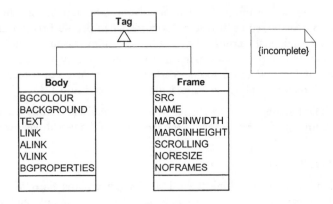

Figure 8.2   *Types of 'Tag'*

is indicated, interestingly enough, by another of the UML extension mechanisms, the 'constraint'. The term '{incomplete}' is actually a type of constraint and tells us that more subclasses of the class 'Tag' do exist, but that they are simply not shown on this particular model.

'Body' is a type of 'Tag' that has the following attributes: 'BGCOLOR' 'BACKGROUND', 'TEXT', 'LINK', 'ALINK' and 'BGPROPERTIES'. These attributes exist in HTML and allow the appearance of each Tag to be customised. These properties will dictate the appearance of the HTML as seen by the person viewing the web page.

'Frame' is a type of 'Tag' that has the following attributes: 'SRC','NAME', 'MARGINWIDTH', 'MARGINHEIGHT', 'SCROLLING', 'NORESIZE' and 'NOFRAMES'.

It should be quite clear that this model could be extended to include all types of 'Tag' and, as the HTML is a finite language, it is not unfeasible to do so.

8.3   Stereotypes

8.3.1   Introduction to stereotypes

One of the reasons why the UML is so flexible is that there are several extensions mechanisms that allow the UML to be customised for a specific application [3, 4, 5]. One such mechanism is known as 'stereotypes'. A stereotype allows new types of UML elements to be defined – but only as long as they are based on (effectively a subclass of) an existing element. This allows the semantics of the UML to be extended, but not the basic structure of the UML. Think of stereotypes as a refinement of existing elements. In reality, stereotypes actually add new elements to the UML meta-model as the UML language itself is being modified. Having said this, however, it is only possible to enhance existing UML elements rather than create new ones from scratch.

These stereotypes represent a distinct usage of the UML, which is usually application-dependent. As a consequence of this application dependency, the models may require an expert in the relevant application domain to interpret the models. This is potentially dangerous as it actually goes against part of the philosophy of the book, that of making communication simpler. However, providing that these stereotypes are well defined and the meaning is assumed by this reader of the UML model, the risks may be minimised.

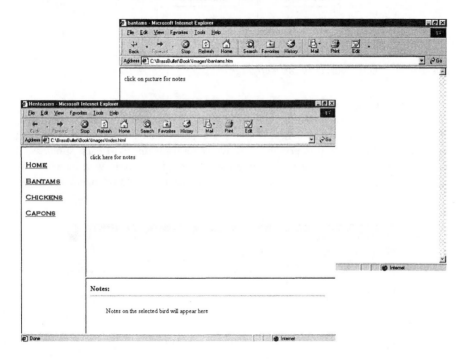

Figure 8.3 *Example (fictitious) website*

Figure 8.3 shows some example web pages that make up a website that will be used as the basis for the explanation of stereotypes.

Consider a very simple example of a few pages on a website called 'www.henteasers.com'. This is, as far as the author is aware, a fictional website and should not be searched for on the world-wide web. There is a main page called 'Bird index' and three further pages that may be accessed from this main page, which are 'Bantams', 'Capons' and 'Chickens'. Each of the three further pages is accessed by selecting from a list of birds in the left frame of 'Bird index'. The information contained in each further page is displayed in the right frame of 'Bird index'. This may be modelled in a number of ways.

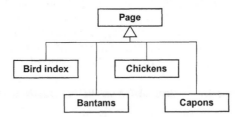

Figure 8.4   *UML model of the web pages*

In order to represent these pages in the UML, the model shown above was drawn up. It can be seen that there are four types of page: 'Bird index', 'Bantams', 'Chickens' and 'Capons'. These classes are shown as subclasses rather than instances of 'Page', as it is possible to have several instances of each subclass. It is perfectly possible for more than one user to access a single web page at any given time, therefore showing instances on this diagram would be incorrect on

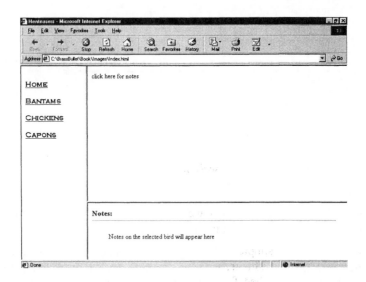

Figure 8.5   *A more complex page*

this model. This is still easy to understand, but does not reflect the true nature of the page 'Bird index', as this page is more complex than the others. Figure 8.5 shows the structure of the main page in more detail.

Upon examination of Figure 8.5, it can be seen that the page 'Bird index' is actually made up of three frames: one for birds, one for pictures and one for facts. These new facts may now be added to the UML model.

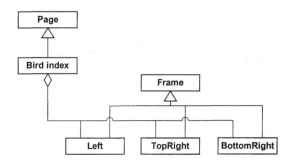

Figure 8.6 *UML model of more complexity*

It can now be seen from the model in Figure 8.6 that 'Bird index' is made up of three types of 'Frame': 'Left', 'TopRight' and 'BottomRight'. Although this model is correct, it is easy to see that the model could soon get out of hand in terms of its complexity, as more and more basic HTML elements are added to the model. This could also be modelled by showing that 'Bird index' is made up of three 'Frame' classes, which has three types: 'Left', 'Right' and 'BottomRight'. However, it is then necessary to explicitly define that there is a one-to-one-to-one relationship between the three classes, as it is not implied from the new model.

This can be avoided by declaring the same basic types as stereotypes, which can then be used as if they were a part of the standard UML.

8.3.2   Defining stereotypes

When a stereotype is defined, we are creating a refined version of a standard UML element. It is not possible to create entirely new UML elements as stereotypes must be based on those already in existence. The starting point, therefore, is to look back at the meta-model of the UML that we would like to refine – in this case, 'Class'.

In Figure 8.7, we are defining several new types of the UML element 'Class' by creating specialisations of it. In this case, there are two specialisations 'Page' and 'Tag', the latter of which is further specialised into 'Body' and 'Frame'. These are the definitions of new stereotypes, which is indicated by showing the word 'stereotype' in chevrons (or 'guillemets'). This pattern is the same as the one introduced in Figure 8.2 to define the HTML language.

What this means is that whenever we see a term in chevrons, such as '<<Frame>>', this is a stereotype that has been defined somewhere as shown in this model. Any class that exhibits the '<< Frame >>' label is assumed to be of the

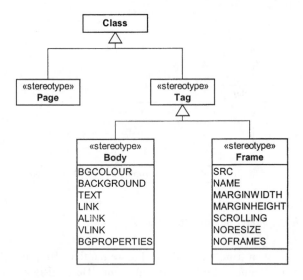

Figure 8.7   *Using stereotypes*

form of the stereotype 'Frame' and takes its structure according to the model above. In this way, it is possible to create stereotypes with their own attributes, operations and relationships that will be inherited by its stereotype classes.

These stereotypes may now be applied to our existing UML model (in this case, the one in Figure 8.6), which is what has been done in the following model.

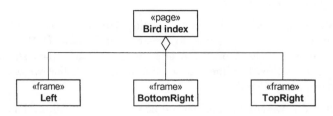

Figure 8.8   *New UML model*

In Figure 8.8, the class 'Bird index' is actually a stereotype called '<<Page>>' from the previous diagram. In the same way, the classes 'Left', 'TopRight' and 'BottomRight' are stereotypes 'Frame'. This is the same as the model shown previously, but is far less complex.

A good way to 'read' diagrams that use stereotypes is to say the words 'that happens to be' whenever they are encountered. Therefore, the model in Figure 8.8 may be read as: there is a class named 'Bird index' that happens to be a '<<page>>' that is made up of three classes: 'Left', which happens to be a '<<frame>>'; 'BottomRight', which that happens to be a '<<frame>>'; and 'TopRight', which happens to be a '<<frame>>'.

8.3.3 The assumption model

There is a danger here, however, as it is important that whoever is looking at the model knows what stereotypes have been created and what properties they possess. This is like making a number of assumptions about a model, and thus a separate model must be created that defines the structure of the stereotypes that are to be used, which has been called the 'assumption model'.

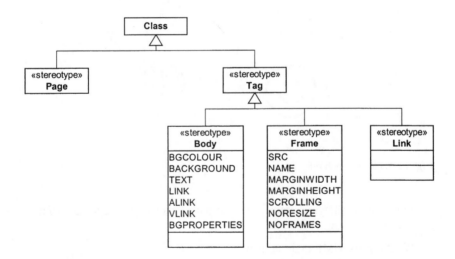

Figure 8.9 *Assumption model*

This assumption model must accompany any models that use the stereotypes defined here – in the same way as maps must have legends.

In the assumption model shown in Figure 8.9, it can be seen that the pattern is identical to that in Figure 8.2; however, this time another stereotype has been added: 'Link'. From the model, it can be seen that there are three special types of 'Tag' called 'Frame', ''Body' and 'Link', all of which happen to be stereotypes.

8.3.4   *Stereotyping symbols*

It is possible to stereotype the actual symbols that represent UML elements to whatever icon is desired. The theory behind this is: that some people will find the UML element symbols too difficult to read; and that stereotyping icons will make it easier for everyone who reads this model. Although both of these points are true, there is a massive danger when doing this as the assumed knowledge that is required in order to read the diagram correctly increases dramatically. If this stereotyping of symbols is taken to extremes, it is almost impossible to guess even the type of UML diagram that the model is representing, let alone what the diagram actually means.

As an example of this, consider the following model and then try to answer the questions posed below.

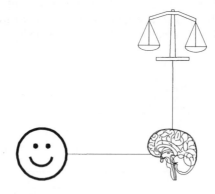

Figure 8.10 *Stereotyped icons*

Figure 8.10 shows an extreme example of stereotyped symbols, but try to answer the following questions:

- What type of UML diagram is it?

- Which UML element does each symbol represent? (Providing that you can answer the first question correctly).

The actual answer to both questions is that it could be almost any diagram! Each UML diagram is made up of a number of symbols that are nodes with arcs that are drawn between them. The model shown here also shows some symbols (although strange) with arcs that join them. Therefore, it is almost impossible to read the diagram without some serious assumptions being made to interpret the new symbols. This may seem like an extreme example, but take a look at some examples shown in UML literature and it will become clear that this model is not as extreme as it first appears.

Having said all this, stereotyping symbols can be very useful and add value to the models; however, extreme caution must be exercised when using them, even more so than with standard stereotypes.

8.4   Summary

In summary, therefore:

- Stereotypes refine existing UML elements. It is not possible to define completely new elements.

- Stereotypes may be used to refine any UML element, not just classes. For example, it may be desirable to create a stereotype of an association called 'invokes' for the web application.

- Whenever stereotypes are used, they must be defined somewhere, such as an assumption model. This acts like a legend to a map.

- It is important to avoid overuse of stereotypes, particularly when it comes to icons, as stereotypes by their very nature assume that the reader has a certain knowledge.

Stereotypes are the first of the three UML extension mechanisms that exist. The next type is 'constraints'.

8.5   Constraints

8.5.1   Introduction

This section looks at the second of the UML extension mechanisms, that of constraints.

Constraints allow the creation of new rules in the UML, which are a semantic restriction represented by a text expression. The constraint itself is shown between two curly brackets and may be represented using an implicit interpretation language, such as a constraint language or a programming language. These may include: OCL (object constraint language), programming languages (such as C++ or Java), pseudo-code or, in some case, natural language.

Constraints are declarative in nature and appear on UML diagrams as restrictions on one or more value of a UML element.

Constraints are limited to being added to only a few types of UML element:

- As an addition to certain UML elements. Although constraints may be associated with any UML element, the use of the OCL limits their use to a subset of UML elements (defined later).

- Attached to a dependency. This is often used, for example, to express a binary relationship between two UML entities, such as an 'XOR' function between two classes.

- Attached to a note. Constraints are often written in notes and associated with UML elements using a dependency relationship.

Constraints are used to express aspects of the UML model that cannot ordinarily be represented using the UML.

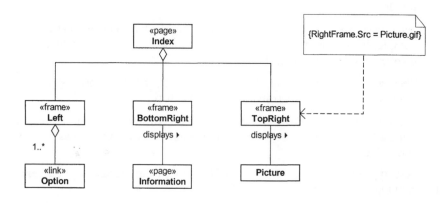

Figure 8.11 *Example constraint*

Figure 8.11 shows how a constraint has been applied to the class 'TopRight' which indicates that the '.Src' value of 'RightFrame' is always set to the value 'Picture.gif'. This ensures that the bottom right-hand frame is always set to a value that refers to a picture page. The constraint itself is written inside a note and its relationship with a class is indicated with a dependency, as defined previously.

8.5.2   Formal constraints – the object constraint language (OCL)

8.5.2.1   Overview

The structure and contents of a constraint can be relatively informal, providing that it makes sense to the reader and adds value to the model. However, for more formal, precise UML modelling there is a bespoke language that is defined as part of the UML standard, known as the object constraint language, or OCL. The language itself is fairly simple yet powerful and can be used to express aspects of a UML model that are not covered by the basic UML syntax. An in-depth discussion concerning the OCL is outside the scope of this book; however, this section gives a very high-level view of the structure and syntax of the OCL. For a complete definition of the OCL and its use, see [6]. This information is presented in the same way as the UML has been presented in previous chapters by using the concept of a meta-model.

The OCL is used to represent constraints that are then attached to particular elements of a UML model. There are three basic types of constraint, as shown in Figure 8.12.

Figure 8.12 shows that there are three types of 'Constraint': 'Invariant', 'Pre-condition' and 'Post-condition'. These will be described in more detail in due course.

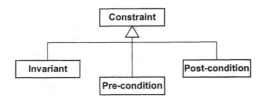

Figure 8.12 *Types of constraint*

8.5.2.2   Invariants

An 'Invariant' describes something that cannot change, such as an attribute value. This is often useful when describing class hierarchies using the specialisation and generalisation relationships, where the difference between two child classes may be the fact that one inherited attribute is always set to a certain value. In order to illustrate this, consider the following example.

Figure 8.13 shows a possible use for an invariant constraint. In this example, the class 'Mammal' is shown as having three attributes – 'number of legs', 'gender' and 'age' – which will be inherited by its two subclasses 'Dolphin' and 'Cat'. Obviously, there are many differences between a dolphin and a cat, but for the purposes of this example we shall consider how we may differentiate between them based on the information contained in the parent class. Consider the attribute 'number of legs', which is applicable to both dolphins and cats. In each case, there is a value assigned to this attribute that (in most cases) does not change. A 'Cat' will always have 'number of legs' set to the value '4' (unless it has been in some sort of accident, in which case set it to between zero and four) whereas the 'Dolphin' will always have its 'number of legs' attribute set to the value zero. This concept of a value never varying is the 'invariant' in the OCL.

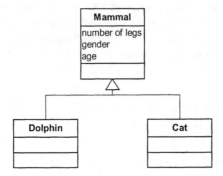

Figure 8.13 *Possible use for a constraint*

8.5.2.3   Pre-conditions and post-conditions

The other two types of constraint are the 'Pre-condition' and 'Post-condition', which are usually associated with operations. In order to understand why pre- and post-conditions occur, one must consider the concept of 'design by contract'. A 'contract' is defined as an unequivocal agreement between two parties in which both parties accept obligation. The two parties, in this case, are the consumer and the supplier, which will be classes in a UML model. The services, or operations, that are available to a consumer are defined by the interface to the class. This is shown visually in Figure 8.14.

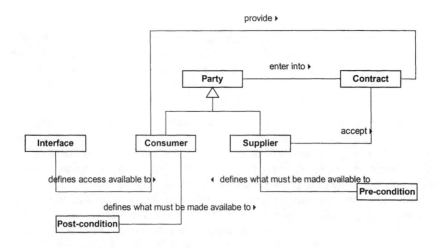

Figure 8.14 *Concepts concerning pre- and post-conditions*

Figure 8.14 shows the concepts concerning pre- and post-conditions related to the previous paragraph. It can be seen in the model, therefore, that:

- The 'Interface' defines the access available to the 'Consumer' of a particular class. This 'Interface' is made up of one or more 'Operation' that will be made publicly available to the outside world. Each 'Operation' is made up of zero or more 'Pre-condition' and zero or more 'Post condition'.

- A 'Pre-condition' states what must be made available to the 'Supplier' when an operation is invoked. To put this another way, the 'Pre-condition' describes the conditions that must be met, including any information that must be made available, in order for the operation to execute correctly.

- A 'Post-condition' states what must be made available to the 'Consumer' once an operation has been executed. To put this another way, the 'Post-condition' describes what is expected as the outcome of the execution of the operation.

These three types of 'Constraint' are represented using OCL expressions. Each OCL expression has the same structure, which is illustrated in Figure 8.15.

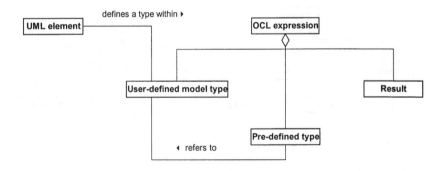

Figure 8.15 *Structure of an OCL expression*

Figure 8.15 shows that an 'OCL expression' is made up of a 'User-defined model type', a 'Pre-defined model type' that refers to the 'User-defined model type', and a 'Result'. The 'UML element' defines a type within a 'User-defined model type' and may be further described using the model below.

Figure 8.16 shows the four basic types of 'UML element' that are used in the OCL: 'Class', 'Use case', Data type' and 'Actor'. It is these UML elements that define the 'User-defined model type'. This relationship effectively allows a constraint to be associated with a UML element by stating that the UML element defines the type required in the OCL expression. Effectively, this means that the constraint can only be associated with the UML element (and any other their sub-elements, such as attributes and operations).

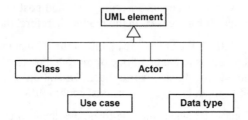

Figure 8.16 *Types of UML element*

The 'Pre-defined types' from Figure 8.15 may also be defined in more detail, as shown in Figure 8.17.

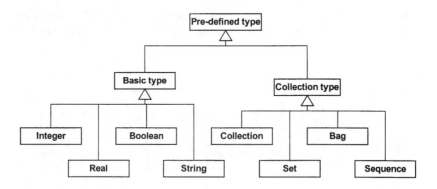

Figure 8.17 *Description of user-defined types*

Figure 8.17 shows that 'Predefined type' has two subclasses: 'Basic type' and 'Collection type'. 'Basic type' has types 'Integer', 'Real', 'Boolean' and 'String', which are the same as their mathematical namesakes. 'Collection type' has four subclasses, which require a little more description:

- 'Collection'. A 'Collection' is a generic name for 'Set', 'Bag' and 'Sequence', which contains their common operations.

- 'Set'. A 'Set' is a type of 'Collection' that contains instances of OCL types. Each instance can only be present once in the collection and the instances are not ordered.

- 'Bag'. A 'Bag' is a type of 'Collection' that contains instances of OCL types. Each instance can be present more than once in the collection and the instances are not ordered.

- 'Sequence' A 'Sequence' is a type of 'Collection' that contains instances of OCL types. Each instance can be present more than once in the collection, but this time the instances are ordered.

Each 'Pre-defined type' has a set of operations associated with it that have a name, an expression and a result type. The model below shows the operations associated with the type 'String'.

Figure 8.18 shows that the 'String' type has seven operations: 'concat', 'size', 'toLower', 'toUpper', 'substring', 'equals' and 'notEquals'.

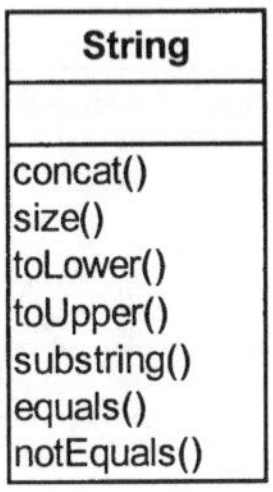

String
concat()
size()
toLower()
toUpper()
substring()
equals()
notEquals()

Figure 8.18 *Operations of 'String'*

Let us now consider some examples of OCL expressions. The constraint expressed visually in Figure 8.13 may be shown in OCL as:

Cat
Number of legs = 4

Where '*Cat*' is the class name and '*number of legs*' the attribute to be constrained. The attribute is of type integer, therefore it may use any of the operations defined on the pre-defined type 'Integer'. The equals sign '=' is one such operation and its value ('Value' from Figure 8.15) is assigned as '4'.

The 'Collection' types deal with single-valued or multi-valued attribute. As an example of this, imagine that we have added a new attribute called 'name' and we would like to define its value as a set of three names (all cats have three names):

Cat
name = Set {name}

Where '*Cat*' is the class name and '*name*' is the attribute to be constrained. The attribute is of type set, therefore it may use any of operations defined on the pre-defined type 'Set'. The value ('Value' from Figure 8.15) of the set is assigned as all values that can be found for '*name*'.

For a complete and far more detailed description of the OCL, see Warner's definitive book on the subject.

8.6   Tagged values

This section looks at the third UML extension mechanism, that of tagged values. Tagged values allow extra information to be added to any UML element and are shown as tag-value pairs. This means that each tagged value is made up of a tag string, a value string and an equals expression.

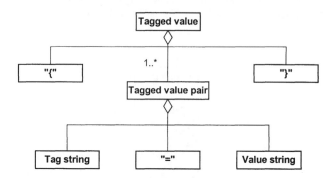

Figure 8.19 *Tagged value syntax*

Figure 8.19 shows that a 'tagged value' is made up of a '{', one or more 'Tagged value pair' and a '}'. Each 'Tagged value pair' is made up of a 'Tag string', a '=' and a 'Value string'.

Tagged values are typically used to show extra information, such as project management information or configuration management information. Tagged values are not part of UML semantics, but are information about the model itself.

Figure 8.20 *Example tagged value*

It can be seen from the model in Figure 8.20 that extra information has been added to two classes, in the form of a tagged value that contains two tagged value pairs. The tag strings are 'Author' and 'Date' and the value strings that define the value for each.

Sometimes it is desirable to define a set of tagged values that may be added to a particular element. In such a case, they should be defined on an assumption model by defining types of 'tag string' as defined in Figure 8.19.

8.7 Conclusions

This chapter has introduced the three basic extension mechanisms that exist in the UML:

- Stereotypes. These are used to define new UML elements based on existing ones. Specialist knowledge is required to understand a model that uses stereotypes, which must be represented on an assumption model.

- Constraints. There are three types of constraint: invariants that restrict the value of a UML element and pre- and post-conditions, that define what happens before and after the execution of an operation.

- Tagged values. The use of tagged values allows extra information such as project management and configuration management details to be added to UML elements.

All these extension mechanisms should be used with caution, as it is very easy to make a diagram more confusing than it might otherwise be. Remember, if you are unsure about whether to use stereotypes or not, ask yourself whether it adds value to the model or in fact detracts from it.

8.8 Further discussion

1. Revisit the new process that was modelled as part of the process modelling exercise in Chapter 6. Consider the high-level structure model and apply stereotypes to it. Where can stereotypes be used here? How will this affect other models?

2. Define the assumption model for your stereotypes from question 1.

3. Find out more information about the OCL and model some of its pre-defined types and their associated operations, as shown in Figure 8.18.

4. Why might you prefer to use OCL over informal constraint modelling, or vice versa?

5. Create a set of tagged values that may be applicable to UML components from any component diagram from Chapter 5. Are there any other UML elements to which your tagged values may be applied?

8.9 References

1 CONALLEN, J.: 'Building web applications with UML' (Addison-Wesley, 1999)

2 BROOKS, F. P.: 'The mythical man-month' (Addison-Wesley Publishing Inc., 1995)

3 RUMBAUGH, J., JACOBSON, I. and BOOCH, G.: 'The unified modelling language reference manual' (Addison Wesley, 1998)

4 BOOCH, G., RUMBAUGH, J. and JACOBSON, I.: 'The unified modelling language user guide' (Addison Wesley, 1998)

5 JACOBSON, I., BOOCH, G. and RUMBAUGH, J.: 'The unified software development process' (Addison Wesley, 1999)

6 WARMER, J. and KLEPPE, A.: 'The object constraint language, precise modelling with the UML' (Addison-Wesley Publishing, 1999)

Tools

"hi, I'm Plenty"
– Plenty O'Toole

9.1 Introduction

This chapter discusses tools that may be used for UML modelling and, in some cases, that may be used for general systems engineering tasks. The main aim of this chapter is to provide information and guidelines that describe how to make an informed decision when choosing a tool.

Tools can be expensive, very expensive! It is therefore absolutely essential to know your requirements when choosing a tool and to make sure that your requirements drive the choice of tool, rather than the other way around.

No tools will be mentioned explicitly in this chapter as the tool market fluctuates very quickly. The functionality of tools changes enormously from one version to another, or from one configuration to another. The marketplace itself is still quite unstable with larger companies swallowing up smaller ones in order to take over their market share or, in a slightly more sinister way, to remove a competitor from the market altogether.

Some companies are quick to dispel their competition with swift statements, while others produce reports and comparison tables between their tools and their competitors'. It is worth remembering where tool vendors actually make their money – is it from the sale of the tools or from the maintenance costs that go along with it? Is it in their interest to provide a tool that is simple and easy to learn when they provide training courses and consultancy to help the way you work with the tool? This may seem like a particularly cynical view, but when spending such potentially large sums of money a healthy dose of cynicism can save a significant amount of money.

This chapter is split into three main sections:

- The first section gives a brief overview of CASE (computer aided/assisted software engineering) tools and discusses what a typical tool can do. Perhaps even more important than this is to realise what a tool cannot do and to have a good picture of what its limitations are.

- The lion's share of this chapter is devoted to providing a number of guidelines to follow, or at least bear in mind, when choosing a tool. The guidelines in this section are based on many years' personal experience dealing with both tools and their vendors and also seeing the result of what making the wrong decision can have on an organisation or project.

- Finally, this chapter winds up with some conclusions concerning tools.

All of these points are designed to be generic so that they may be applied to the selection of any CASE tool whatsoever, rather than trying to compare specific tools on the market. Clearly, the latter approach would render this chapter obsolete even before the book was published, bearing in mind the speed at which tools change and the rapid growth in the number of CASE tools on the market.

9.2   Introduction to tools

The tools that are discussed in this chapter may be split into two broad categories – CASE tools and systems engineering tools – although generally speaking the guidelines in this section are equally applicable to both.

The acronym CASE stands for either 'Computer Aided Software Engineering tools' or 'Computer Assisted Software Engineering tools'. Although both definitions are correct, the true definition is lost in the mists of time and nobody is really sure which of the two definitions is the true one. CASE tools are generally associated with a particular modelling language, method or methodology. Obviously, the tools that this book is aimed at are the UML-based CASE tools, although the guidelines are applicable to any type of CASE tool.

Systems engineering tools are often similar to CASE tools, but they do not generally adhere to a particular approach. Their functionality tends to be aimed at a higher level – dealing with process-oriented functions rather than modelling *per se*.

Some of today's tools actually do both, or may be integrated into a 'suite' of tools that are purported to make life far easier. Beware of words such as 'seamless', 'transparent' and 'single-stop solution' when considering such tools as they are concepts that are much vaunted yet rarely exist in real life.

9.3 Selecting a tool

9.3.1 What tools can do

Before entering the world of CASE tools, it is very important to take a reality check and to sum up what these tools are capable of. Generally, CASE tools can help a project in three ways:

- The main use of CASE tools is to help the people who want to model to use the language in a consistent and constructive way. The tool should aid a newcomer, yet at the same time not hinder an experienced user. Most CASE tools will make modelling easier in that: models will be quick and efficient to produce, models will look good (hence aiding communication) and they will help to store the diagrams in the model in a simple and efficient manner. As a basic rule of thumb, imagine creating UML diagrams using a simple drawing package and then make comparisons based on its usage compared to a CASE tool. This is the benchmark that is used throughout this section.

- The second way that CASE tools can make life easier is, like all computer applications, to take the drudgery out of a particular task, in this case by helping with checking and verification. This becomes very important with regards to syntax and semantics and can, theoretically, be used as part of a quality control process as a first step towards assessing models.

- The third major use of CASE tools is to help with documentation. In an ideal world, documentation should be fully automated and a simple keystroke should automatically produce any report that is required. The reality of this is somewhat different, but this will be discussed further in due course.

CASE tools do have many other functions but these may be grouped into the 'bells and whistles' category from a systems engineering viewpoint. Code generation is purposely omitted here as a main feature as it is limited to software engineering, rather than the wider scope of systems engineering.

Systems engineering tools generally work at a higher level, being more concerned with process activities than actual modelling. The typical functionality of a systems engineering tool includes:

- Traceability throughout life cycle phases. Being able to trace individual requirements forward through the project life cycle is very important. The opposite is also essential, for example, starting at one point in the design and trying to trace back to its originating requirement. This is the main focus of many systems engineering tools.

- Information management, where information pertaining to a project (that should be based on a process) can be stored in a central repository across all phases of the systems life cycle. This will include managing the implementation model if the approach suggested in Chapter 6 is followed.

- Process definition and help. This is potentially the most useful aspect yet also potentially the most dangerous aspect of tool functionality. It is stated (over and over) in this book that the process must drive the tool and not the other way around. By letting a tool effectively organise your process, it is relatively easy to fall into this trap.

This chapter will focus on CASE tool functionality with regard to the UML, rather than general systems engineering functionality. It should not be too difficult, however, to extract a few rules from the ones presented here that would be applicable to almost any tool.

9.3.2 What tools cannot do

The previous section introduced some high-level features of what CASE tools can do, and now this section looks at the opposite of this, which is what CASE tools cannot do. The non-functionality of CASE tools may be summarised as follows:

- Validation. Despite what people may say, there is no inherent mechanism for current tools to validate designs. Designs may be verified using the mechanisms mentioned previously, but there is no way to validate whether these designs actually meet their original requirements or not.

- Ensure correct models. There is no way that a model may be automatically checked to ensure that the model makes sense. Using an example that has been used previously in this book, there is nothing to stop someone creating a model that reads 'cat' chases 'dog', which in our stereotypical world cannot happen. A tool can check the syntax of the model, but cannot guarantee that the overall model is correct.

- Another function that CASE tools cannot perform is to implement a generic process. Many tools implement a process, but this will probably not be the process that a given organisation would want to implement. This can be very dangerous, as adopting a proprietary process can lock an organisation into a single vendor's product, which can turn out to be very expensive. This is a situation that has been encountered all too often in the past. The problem starts when an organisation buys a proprietary tool which implements a proprietary process. The organisation is then convinced that this process is an appropriate approach for their particular set up, without any sort of analysis or assessment of the organisation. Quite often, this buy-in will occur when the organisation is shown a ream of case studies and an example of how the process works that has been implemented using the tool. The next step is that the organisation, when it comes down to implementing the process on a real project, does not have the practical knowledge needed to use the process properly, and thus calls in consultants from the company that produces the tool. The next step is that training courses are required in order to achieve the full benefits of the tools and the process. A few months later, the price of all consultancy and training goes up significantly, but the organisation is now locked into using a particular tool and approach and has invested too much time and money to pull out at this stage. Another eventuality is that the tool vendor then goes out of business or is bought up by a larger company, effectively reducing the resources required for the project.

This section may seem rather negative, but it is important to approach assessing any tool with both eyes wide open.

9.3.3 Typical CASE tool capabilities

Different tools offer different capabilities, but there is a core set of functionality that is offered by many of the tools. Many specialist capabilities will not be discussed here, such as real-time extensions and language-specific constructs.

The functionality of CASE tools has been grouped into six main categories that will, in the case of some tools, describe all or more than their available functionality and, in the case of some others, very little of their functionality:

- Drawing diagrams, this is the most fundamental of functions for CASE tools. After all, no diagrams – no model!

- Verifying syntax, which is the 'spell-checker' of the CASE tool world. If your models are inconsistent they will translate in real terms to an unrealisable product.

- Software-specific features, such as code generation and reverse engineering. Many CASE tools now have these features to some degree. They may appear in different versions of the individual tool. Code generation takes models and automatically creates code stubs from them. Reverse engineering is the same as the last point but in reverse. The tool takes the code and creates a model from it.

- Report generation. This feature is useful in providing a complete account of what is in your models.

- Import/export with other packages. CASE tools will usually be introduced into an existing working environment where other tools will already be established and accepted as standard working practice. In reality, any CASE tool will need to be able to import from and export to other tools and packages and, in an ideal world, this integration should be seamless.

This list covers basic functionality, but the way in which each of the facilities can be assessed can be extended to cover any other functionality.

9.3.3.1 Drawing diagrams

Perhaps the main reason to buy a CASE tool is to help with drawing the actual diagrams themselves. While it may be practical to create initial models using a paper and pencil or a white board with sticky notes, there comes a point in every project where it becomes too impractical to continue using a hard medium, due to the constant changes that must be made to each model, particularly in the early stages of a project. These constant changes are due, in part, to the iterative nature of UML development.

It may also be tempting to use a standard drawing package to create UML diagrams – after all, one of the requirements that was used when creating the UML itself was to make it simple to implement on an automated tool. Any drawing package, no matter how basic, is capable of creating any UML diagram, which must be seriously borne in mind when looking at some of the more expensive CASE tools that are around today. One of the problems that existed with the older modelling techniques was that the symbols chosen were a little esoteric at times. Anyone who has ever tried to draw a Booch class, represented by a 'cloud' shape, will know the problem. Anyone who has ever tried to draw a 'dashed' cloud either by hand or with a tool will also know frustration!

Some sophisticated drawing packages allow the creation of special templates that allow specific shapes to be created automatically, which means that it is possible to completely configure a CASE tool out of a simple drawing package! This may seem completely impractical, but almost all of the tools around at the moment do not meet the UML 100 per cent when it comes to drawing the diagrams! It is important, therefore, to ensure that you are getting more out of the CASE tool than just a bespoke drawing package, especially bearing in mind the difference in cost between a simple drawing package and a full-blown CASE tool.

When it comes to assessing the drawing capabilities of a CASE tool, the first and most obvious question (indeed, so obvious that most people do not bother to find it out) is to find out how many of the nine diagrams the tool will allow to be created and whether or not the diagrams actually match the UML standard itself! Whenever this question is raised in a group discussion, the initial reaction is usually one of scorn and disbelief that anyone could produce a tool that does not match the UML. Some tools have the capability to model as few as four of the nine diagrams, while others offer a facility to allow up to 15 different diagrams to be created – all of which are allegedly UML diagrams.

Once it has been established which of the nine diagrams can be created, the next step is to see how each of the diagrams relates to the UML. Some diagrams follow the UML precisely, while others only bear a passing resemblance to their UML counterparts.

It is quite bizarre that so many tools do not measure up to the UML, particularly bearing in mind that the UML is very cleanly defined and has a single source document as its reference. The following list suggests reasons why some CASE tools may not be as UML-compliant as perhaps they should:

- Many CASE tools are an evolution of existing, or legacy, tools. The methodologies behind these legacy tools often provided an input to the UML, and thus many vendors may just assume that the diagrams used (for example,

OMT) are the same as the diagrams in the UML. Unfortunately, these diagrams tend to be slightly different from their original language.

- Many companies are reluctant to throw away existing resources, so they may deem that it makes more business sense to re-use an older tool, rather than to re-develop it properly for a new market.

- It is also highly possible that some CASE tools are sometimes developed by people who do not actually use the technique themselves An outsider's view point of what may be useful may be completely different to what a practising UML specialist may deem as being useful.

- It is often the case that tool companies are looking for a market 'edge' and will focus their tool towards a specific market, such as real-time systems or safety-critical systems. In addition, the tools may be aimed purely at a single target, such as a particular programming language. Sometimes, this specialisation of a tool can be to the detriment of its compatibility with the UML.

Another aspect of using a CASE tool that is worth considering is how the tool navigates between the different diagrams. Some tools may allow any diagram to be created outright, while others may insist that one diagram exists before another may be created. As an example of this, consider the strong relationships between the nine diagrams that have been discussed so far in this book. Statecharts, it has been stated, should be associated with classes or use cases; therefore, some tools will only allow a statechart to be created if it has a class or use case associated with it. Another example is where a tool has adopted a particular process that insists on the diagrams being created in a strict, pre-defined order such as: use cases first, interaction diagrams (collaboration and/or sequence diagrams) and then class diagrams. Another tool may insist, for example, that class diagrams are the first diagram to be created and all other diagrams should relate back to this.

On a more practical note, it is worth having a test model in mind that uses a few types of diagram that can be used to test the compatibility of a tool with the UML. A good example of this is the chess example that was used throughout Chapters 3 and 4 of this book, which is simple, yet uses all nine of the UML diagrams. Another useful thing to do is to enter the model yourself, rather than relying on a demonstrator to do it for you. This approach will provide a better indication of how easy the tool is to use and may very well throw up some quirks of the tool itself.

In summary, therefore, the following points should be looked at when assessing the drawing capabilities of a tool:

- Does the tool allow all nine UML diagrams to be created and are the correct names used for each diagram?

- Do each of the diagrams that may be created actually match the syntax that is laid out in the UML standard?

- How does the tool allow navigation between different types of diagram?

- Have a sample model that may be input into the tool by way of a simple test. Choose something simple that will not take long to input and put the tool through its paces (such as the chess example in this book).

Remember that drawing is fundamental to UML and thus it is important that the tool can perform well.

9.3.3.2   *Verification*

Every CASE tool on the market will offer some form of verification that can be automatically performed on the UML models. Invariably, this turns out to be some sort of syntax checking that may be carried out on diagrams and consistency checks between diagrams. These checking facilities may be automatic, where the syntax is checked as the model is entered, or they may have to be invoked explicitly. The actual level of checking varies enormously and the following list gives some indication of the range of the verification facilities of tools, based on the relationships between all types of UML diagrams:

- One of the simplest checks is to ensure that there are no name clashes in a model. A simple example of this may be to check that no two classes have the same name. This is a feature that is often automatic and, in many cases, the tool may actually forbid a name clash, rather than just providing a warning.

- Another popular check is to ensure that send and receive events are consistent in some way. There are many ways to ensure that events are consistent that range from simply checking that each send event that exists has a related receive event on the model, to ensuring that arguments match up that are passed between objects.

- Some tools will ensure that any activities that are used on a statechart are chosen from a list of operations from their associated class diagram. In fact, some tools will actually add any activities that do not exist as operations to their associated class diagram.

As can be seen from this list, the functionality of checking facilities varies, but there is a more subtle danger with using a tool's checking facilities and that is how the checks are actually performed. This may seem trivial, but if the checking facilities are being used as part of a quality system, or as part of a defined process that will, in some way, demonstrate the correctness of a model, it is absolutely essential that the underlying rules behind each check are explicitly defined. This is a problem that rears its ugly head in two of the other aspects of the assessment criteria laid out in this chapter.

The final point to bear in mind with verification is what actually happens to the results. Are they displayed on the screen or are they saved in a file somewhere? Will they still exist once the display window has been closed down or will they be deleted, never to be seen again? It may even be desirable to associate them with a particular version of a model (which relates to one of the other points here: version control).

The following list summarises the basic assessment criteria for model verification facilities in a tool:

- What verification facilities are offered by the tool?

- Are these verification facilities defined in some way, or are they non-deterministic?

- How are the results displayed and stored?

If the tool cannot offer adequate checking facilities, what exactly is it offering over and above a standard drawing package?

9.3.3.3 *Software-related features*

It must be remembered that the UML's background is primarily derived from software-based methods and methodologies, rather than generic systems modelling; therefore, almost every CASE tool will come with some type of code generation function.

Most major languages are supported by CASE tools and some offer support for more than one language. The number of languages that are supported, however, will often have quite an impact on the price paid for the tool. For example, some tools will have different editions, depending on what you want them to do. They have names such as 'basic', 'modeller', 'professional', 'designer' and 'enterprise'. Some of these editions will not allow any code generation, whereas others will allow a single language, or full support for all languages that are available.

It is also worth checking how the tool will actually integrate with the target language's development environment – or, indeed, whether it will at all. Some tools will interact with a fourth-generation programming language by controlling it in some way, whereas other tools will simply create text files that need to be explicitly imported into a development environment.

Perhaps the most fundamental aspect of code generation to be looked at is whether it actually saves any time and effort. This may seem like a strange statement to make, but many people use code generation once and never touch it again. It is worth asking what code is actually created. In most cases, this will almost certainly be definitions of, for example, classes, attributes and operations, along with stubs for each operation. Although useful, is this worth the extra money in licence costs to have the ability to create code stubs? Which diagrams are actually used for the code generation? The main contenders for this are those of the static model, as, by their very nature, they represent a snapshot of the software that is relatively simple to implement. Now think of the behavioural model and think what aspects of this model are used, if any. Also, remember that any information that is generated in the code must exist somewhere in the model.

Imagine, for example, that the code that is generated contains information to define attribute types, default value, visibility etc, and this information must be entered (typed) into the model. At the end of the day, how much time is this saving, if the level of detail of the model is so low that the model is almost code itself?

There is a more dangerous aspect to code generation, which is the same point that was raised in the previous section concerning verification. How exactly is the code generated? What are the rules that are followed when going from a class diagram to a piece of code? If this process of transforming a design into code is not fully defined, how far can it be trusted? History and fiction has taught us that we simply cannot trust any system that is not fully deterministic. In reality, this is impossible on all but the most trivial of systems, but that does not mean that we can be complacent by allowing an unknown, undefined process to create (potentially) safety-critical applications. Remember the disaster that occurred in the Hollywood movie *Westworld*, where human-like android slaves turned into psychotic killing machines simply because they had the capability to change their own programming, which put them out of direct control of their human creators.

Code generation can be very useful and very productive, but, if not assessed and understood properly, it can also be a complete waste of time, money and disk space.

The following criteria should be borne in mind when assessing code generation facilities in a tool:

- Do you actually want code generation?

- What are the extra costs for code generation functions compared, for example, to the basic modeller-only edition of the tool?

- Which languages are supported by the tool and which development environments? In addition, which manufacturers and exact version numbers are supported?

- How much time and effort is actually being saved by using the code generation functions? Put it to the test before spending any money.

- What code exactly is generated? Is it just code stubs? Which diagrams are used to generate the code? Some tools claim 100 per cent code generation, which is quite a bold claim, but what does this mean exactly? In some cases, this means that 100 per cent of the model is used to generate the code rather than being able to generate 100 per cent of the actual final code.

- How, exactly, (by what process) is the code generated? Are the rules defined anywhere or are they non-deterministic?

These guidelines are applicable to forward engineering only and not reverse engineering, which is covered in the remainder of this section.

Reverse engineering is, unsurprisingly, the opposite of forward engineering. If forward engineering is going from designs to code, then reverse engineering is taking existing code and generating designs from it. This may be a particularly attractive function, particularly where a great deal of legacy code exists that is difficult to make sense of.

Clearly, the output from such a function will be the models themselves, so it is important to stop and think for a moment about what types of model are created. First, take this literally and ask which of the nine diagrams are created from the code. Perhaps the most obvious diagram is the class diagram, but what about the others? It is only possible to reverse engineer between the implementation (code) phase and the low-level design phase, so how much practical use will the tool be?

Much of the focus of this book is how to communicate effectively by creating simple, clear and easyto understand models. Much of the art and skill of UML modelling is getting these models at an appropriate level of abstraction and ensuring that they have a clear connection to reality. The connection to reality should not really be a problem with regard to the reverse engineering tool, as it is taking the reality and creating diagrams from it. But what about the level of abstraction of the models that are created? What about similar views of the same system? How can an automated tool possibly define, for example, sensible scenarios based on the code? The simple answer is, of course, that a tool cannot! It is well worth having a look at the quality of diagrams that are output from the tool, as in some cases they contain very large classes that are full of attributes and operations at a very low level of abstraction, which, for all intents and purposes, are useless.

An obvious aspect of reverse engineering to look at is which languages, environments and version of these are supported by the tool? Choosing an incorrect version number or the right language from the wrong manufacturer can be very costly.

The final point is a recurring one, which is, how are the models generated and what rules are followed to generate the models? Again, it is a question of having a deterministic system or a non-deterministic one.

The assessment criteria for reverse engineering may, therefore, be summed up as follows:

- Which of the nine UML diagrams are generated from the code?

- How are the diagrams generated from the code – are the rules defined?

- Which languages, environments and their associated versions are supported?

- How useful is this function?

Above all, remember that reverse engineering will not be a feature that is wanted or needed by everyone, and is it worth assessing exactly how much it would be used before using it as a basis for tool selection.

9.3.3.4 *Report generation*

One of the potentially most useful functions offered by CASE tools is report generation. Report generation is particularly attractive when demonstrations are shown where a complete phase deliverable that is produced at the touch of a single key.

The first criterion is to look at the output of the report generation and ask whether the tool produces its own files or whether it produces files that are compatible with another product. Many tools will offer both facilities: a simple text-only output that may be generated quickly and simply from the data dictionary of the tool, or a complex file that may be used by a proprietary word-processing package. In the case of a simple text output, any text editor will suffice, ranging from the most sophisticated office product to the simplest of line editors. However, what most people will want to see is a completely formatted, beautifully laid-out document that can be printed out directly from a favourite word-processing tool. This is a great idea, but the format of the document will be whatever the tool vendor thinks the format should be. This may be avoided by some type of editing function that will allow templates to be devised to create bespoke documents. The amount of effort involved in this may vary, however, from a simple 'tick the box' approach, to having to have intimate knowledge of a particular programming language or script language that may be required to create such templates. This is particularly important when it comes to following, and thus complying with, a process. It would be very useful and efficient to have document templates defined that are compatible with a defined process, which would be very beneficial indeed to any project.

Another related note that is worth considering is which word-processing packages is the tool compatible with and is another licence required? This is one of the ways that extra costs can be sneaked on to the bill without anyone realising it. This is not much of a problem with PC-type systems, where most word processors seem to be converging in terms of format and compatibility, or that are relatively inexpensive anyway. However, consider, for example, Unix-based systems where the software (for dark reasons known only to software companies) seems to cost orders of magnitude more than their PC counterparts.

If the CASE tool is interacting with another package, it is worth considering how efficient the interaction between the two is. In some cases, there is no interaction whatsoever, as a file is created which must then be opened from the appropriate word-processing package. Another option is that the word-processing package is actually controlled directly from the CASE tool. This may have serious effects on the efficiency of the system. In one example, a particular CASE tool took over ten hours to generate a report from a model consisting of less than 50 classes! This turned out to be because the tool was, quite literally, cutting and pasting each diagram to a clipboard and, from there, copying it into the actual word-processing package.

The final point to bear in mind is the old chestnut of how the reports are generated. One particularly ambitious CASE tool was able to not only put the diagrams from the tool into the word-processing package, but also to generate English sentences from the model elements that could be used to explain the model to readers of the report. Unfortunately for the CASE tool company, the aggregation and specialisation interpretations had been confused. Therefore, whenever the model said aggregation, the accompanying text read 'has types' and the specialisation relationship read 'is made up of'. Of course, this could lead to considerable confusion especially if this error was carried over into the code generation or reverse engineering functions of the tool.

The assessment criteria for report generation may, therefore, be summarised in the following list:

- Does the tool use text-only outputs or is the output compatible with a proprietary word-processing tool?

- Are extra licences required to see the output of the report generation and, if so, how much extra is this?

- How easy or difficult is it to create bespoke templates for special report formatting?

- How efficient is the report generation in terms of time?

- What are the rules for report generation and how are they defined?

Report generation can be a very quick and efficient solution to many problems and may benefit a project enormously. However, as with all aspects of CASE tools, caution must be exercised when assessing this aspect of the tool.

9.3.3.5 *Import/export with other packages*

In many real-life scenarios, the CASE tool that is chosen will probably have to integrate with an established working environment. This will entail the CASE tool output being used by other tools that may be used, for example, for aspects of project management. Some CASE tools vendors, sensing this, have geared their product towards becoming a single component of a complete solution that means that the actual UML CASE tool is part of an overall suite of tools that will perform every single task that your organisation could possibly want, which is good.

Some tools will offer extra functionality bundled into the CASE tool itself. This is all well and good, but will it fit into the existing working environment, or will it actually force people to work in a way that the tool wants people to? Extreme caution must be exercised here, as it is another example of the tool driving the process, rather than the other way around.

Some of the functions that may be worth thinking about are listed below. Bear in mind, however, that some of these may be included in the CASE tool or may be available as part of the same suite, or may even (the most likely case) be tools that are already being used within the organisation.

- Project management tools. The CASE tool may interact with such tools in order to make a link between life cycle phases and aspects of project management.

- Configuration management and change control. This is essential for any type of real project where the evolution of models must be recorded. This is achieved, in reality, by baselining and version control.

- Requirements management tools. With requirements driving the whole project and defining the basic quality of a project, it is important that they are well defined and well managed.

- Life cycle support tools. Some tools will help with the development process and thus the life cycle of a project. Again, this may tie both project management and engineering activities together, but the tool must be flexible enough to allow any process to be adopted, rather than simply some proprietary process that is put forward by the tool.

- Animation tools. Some tools will animate different aspects of models. This may range from highlighting states that are active as a system is being executed, to executing 'what if?' situations in the form of scenarios.

- Office suite packages. Some packages are now 'fully integrated' into a complete office suite so that there is no need to ever look at another tool, as this one will do absolutely everything that anyone could possibly want. Silver bullets, indeed!

Integration with your existing environment may well be the crucial factor when buying a tool. It is important, therefore, to take stock of the current working environment before choosing a tool.

9.3.4 Business considerations

Overall, it is very important to make an informed decision when deciding which CASE tool to buy. CASE tools can be very expensive and take up quite a percentage of the total projects costs, so it is imperative that one is chosen wisely. The following list contains a number of very high-level questions that you should be asking when considering buying a CASE tool:

- Will the tool be used on all projects, or just one? Can the cost, therefore, be spread among the whole organisation? Will the organisation be adopting a single tool across the whole organisation, or will the current market be assessed and different tools chosen where appropriate? Clearly, there are pros and cons with each approach, but 'which is the most appropriate?' should be the main question.

- What will the total cost of the tool be in terms of the number of licences and the maintenance costs? The maintenance costs for a CASE tool when viewed over a five-year period can easily cost as much as the tool did in the first place. Bear this in mind and consider the whole life of the project, rather than the initial outlay for the tool.

- What other packages are required to use the tool effectively? This may include simple, inexpensive packages, such as word processors, to extra licences for a third-party database.

- What functionality will you actually use? If you have used the assessment criteria outlined previously in this chapter, it may be that you have come to the conclusion that all is required is the basic drawing functionality of the tool. In such a case, there is little point spending extra money on code generators that will never be used, when the basic package or a simple drawing package will suffice. If, on the other hand, you feel that you would make extensive use of these extra functions, it would be well worth spending extra money and buying a more complete tool than a simple drawing package.

- What level of support do the vendor's offer, in terms of upgrades to versions, documentation and on-line or telephone support? This can be critical, as having to bring in the vendor's own consultants for the day can be very expensive indeed.

- Does your process drive the tool? (This may have been mentioned previously in this book!)

Above all, like any type of project, it is vital that you know your requirements for the tool before talking to any vendors. Draw up a list of requirements and stick to it.

9.4 Conclusions

This chapter has introduced some generic criteria for assessing the suitability of a CASE tool. Although the tone has generally been quite negative, this is not necessarily the case. The right tool can save you time and effort and hence money, and it can make life far easier. By the same token, however, the wrong tool will cost you time and effort and hence money! The overall suggestion here is one of caution, rather than a single negative view.

Therefore, to summarise the chapter, bear the following points in mind:

- Draw up a list of requirements for the tool. Bear in mind that the process that you follow should drive the tool and not the other way around, as this can turn out to be very expensive.

- Check how many of the nine UML diagrams the tool can create and whether they are correct or not. A great help in determining this is to have a simple model at hand that can be quickly entered into the tool. An example of this is the chess example that was introduced in Chapters 3 and 4 of this book. This will not only prove which diagrams can be created and how accurate they are, but will also give an idea about the use of tool and navigation between diagrams. Do not buy a tool on the basis of an example, pre-entered model, but try it for yourself!

- Check the licensing agreements fully and read all the small print. It is essential to buy the right edition of the tool and to be sure that you have an appropriate level of support from the vendors.

- Shop around between different vendors and obtain independent advice from existing users, as any company that refuses to give you the contact details of an existing user may have something to hide. Remember that you are the customer and should rightly have the upper hand in any negotiations.

One final point to remember is that CASE tools are not magic, but they can help out. Always remember that the output of these tools is only as good as the engineer behind the tool.

9.5   Further discussion

1. Draw up a list of requirements for a CASE tool that would suit your exact needs. Make this as fanciful as possible but always ensure that there is a need behind each wish on the list. Now create a UML model that represents this list of requirements. As a starting point, consider either a use case model (both business and system contexts will be important) and also a class diagram that will represent the tool itself.

2. Obtain demonstration copies of two or more different CASE tools and compare them with one another. What can either offer that the other does not? What do they both offer and, hence, which is the better or best?

3. Compare each tool with the model created in discussion point 1.

4. Take the chess example from this book and enter it into any CASE tool. Can it model 100 per cent of the UML syntax used in the model? Remember that this book only introduced approximately 15 per cent of the total UML syntax.

Perform a quick Internet search and see how many free CASE tools are available for download. How do these relate to the commercial tools?

Index

Bibliography

Systems engineering references

These references are solely concerned with systems engineering.

BIGNELL, V., and FORTUNE, J.: 'Understanding systems failures' (Open University Press, 1984)

COLLINS, T., and BICKNELL, D.: 'Crash, ten easy ways to avoid a computer disaster' (Simon and Schuster, 1997)

LEVESON, N. G.: 'Safeware, system safety and computers' (Addison-Wesley Publishing Inc., 1995)

LEWIN, R.: 'The major new theory that unifies all sciences, complexity, life on the edge of chaos' (Phoenix Paperbacks, 1995)

O'CONNOR, J., and McDERMOTT, I.: 'The art of systems thinking, essential skills for creativity and problem solving' (Thorsons, London, 1997)

SKIDMORE, S.: 'Introducing systems design' (Blackwell Publishing, 1996)

STEVENS, R., BROOK, P., JACKSON, K., and ARNOLD, S.: 'Systems engineering, coping with complexity' (Prentice-Hall Europe, 1998)

Software-related references with systems content

The following references have a heavy systems content, despite the fact that they are primarily aimed at the software market.

AYRES, R.: 'The essence of professional issues in computing' (Prentice-Hall, 1999)

BAINBRIDGE, D.: 'Computer law' (Pitman Publishing, 1996)

BROOKS, F. P.: 'The mythical man-month' (Addison-Wesley Publishing Inc., 1995)

DAVIES, A. M.: 'Software requirements: analysis and specification' (Prentice-Hall International Editions, 1990)

FLOWERS, S.: 'Software failure: management failure, amazing stories and cautionary tales' (John Wiley and Sons Ltd, 1997)

JACKSON, M.: 'Software requirements and specifications' (Addison-Wesley Publishing Inc, 1995)

SCHACH, S. R.: 'Classical and object-oriented software engineering' (Irwin, 1996)

Software engineering references

The following references are concerned only with software engineering and are included here for completeness sake.

BOOCH, G.: 'Object-oriented analysis and design, with applications' (The Benjamin/Cummings Publishing Company Inc., 1994)

COAD, P and YOURDON, E.: 'Object-oriented analysis' (Prentice-Hall, Englewood Cliffs NJ: Yourdon Press, 1991)

GRAHAM, I.: 'Object oriented methods' (Addison-Wesley Publishing Inc, 1994)

JACOBSON, I.: 'Object-oriented software engineering, a use case-driven approach' (Addison-Wesley Publishing Inc, 1992)

RUMBAUGH, J., BLAHA, M., PREMERLANI, W., EDDY, F. and LORENSON, W.: 'Object-oriented modelling and design' (Prentice-Hall, Englewood Cliffs, NJ, 1991)

SCHACH, S. R.: 'Software engineering with Java' (McGraw-Hill International editions, 1997)

SCHNEIDER, G. and WINTERS, J. P.: 'Applying use cases – a practical guide' (Addison Wesley Publishing Inc, 1998)

SHLAER, S., and MELLOR, S. J.: 'Object-oriented systems analysis – modelling the world in data' (Prentice-Hall, Englewood Cliffs NJ: Yourdon Press, 1988)

WARD, P. T. and MELLOR, S.: 'Structured development for real-time systems' (Yourdon Press, New York, 1985)

WIRFS-BROCK, R., Wilkerson, B. and Wiener, L.: 'Designing object-oriented software' (Prentice-Hall, 1990)

YOURDON, E. and CONSTANTINE, L. L.: 'Structured design: fundamentals of a discipline of computer program and systems design' (Prentice-Hall, Englewood Cliffs, NJ, 1979)

UML references

These references are some of the books (the number is increasing constantly) that are concerned with the UML.

BOOCH, G., RUMBAUGH, J. and JACOBSON, I.: 'The unified modelling language user guide' (Addison-Wesley Publishing Inc, 1999)

CANTOR, M. R.: 'Object-oriented project management with UML' (John Wiley and Sons Inc., 1998)

CONALLEN, J.: 'Building web applications with UML' (Addison-Wesley Publishing Inc, 1999)

DOUGLASS, B. P.: 'Real-time UML' (Addison Wesley Publishing Inc, 1998)

ERIKSSON, H. and PENKER, M.: 'UML Toolkit' (John Wiley and Sons Inc., 1998)

FOWLER, M. with SCOTT, K.: 'UML distilled (applying the standard object modelling language)' (Addison Wesley Publishing Inc, 1997)

LARMAN, G.: 'Applying UML and patterns – an introduction to object-oriented analysis and design' (Prentice-Hall Inc., 1998)

LEE, R. C. and TEPFENHART, W. M.: 'UML and C++, a practical guide to object-oriented development' (Prentice-Hall, 1997)

MULLER, P.: 'Instant UML' (Wrox Press Ltd, 1997)

ROYCE, W.: 'Software Project Management – A unified framework' (Addison Wesley Publishing Inc, 1998)

SCHMULLER, J.: 'SAMS teach yourself UML in 24 hours' (SAMS, 1999)

SI ALHIR, S.: 'UML in a nutshell – a desktop quick reference' (O'Reilly and Associates Inc., 1998)

STEVENS, P. and POOLEY, R.: 'Using UML' (Addison Wesley Publishing Inc, 1999)

RADER, J. R.: 'Advanced software design techniques' (Petrocelli Books, 1978)
RUMBAUGH, J., JACOBSON, I. and BOOCH, G.: 'The unified modelling language reference manual' (Addison Wesley Publishing Inc, 1998)

STEVENS, P. and POOLEY, R.: 'Using UML' (Addison Wesley Publishing Inc, 1999)

Standards

The following are references either to actual standards or to books that describe standards in detail.

EL EMAM, K., DROUIN, J. N. and MELO, W.: 'SPICE, the theory and practice of software process improvement and capability determination' (IEEE Computer Society, 1998)

ELECTRONIC INDUSTRIES ASSOCIATION: 'EIA 632, Processes for engineering a system'

ELECTRONIC INDUSTRIES ASSOCIATION: 'EIA 731, Systems engineering capability model'

HOYLE, D.: 'ISO 9000, pocket guide' (Butterworth Heinemann, 1998)

INSTITUTE OF ELECTRICAL AND ELECTRONIC ENGINEERS: 'IEEE 1220, Application and management of the systems engineering process'

INTERNATIONAL ELECTROTECHNICAL COMMISSION: 'IEC 61508, Functional safety of electrical/electronic/programmable electronic safety-related systems'

INTERNATIONAL STANDARDS ORGANISATION: 'ISO 12207, Software lifecycle processes'

INTERNATIONAL STANDARDS ORGANISATION: 'ISO 15288, Lifecycle management – system life cycle processes'

INTERNATIONAL STANDARDS ORGANISATION: 'ISO 15504, Software process improvement and capability determination (SPICE)'

INTERNATIONAL STANDARDS ORGANISATION: 'ISO 9000-3, Guidelines for the application of 9001 to the development supply and maintenance of software'

INTERNATIONAL STANDARDS ORGANISATION: 'ISO 9001, Model for quality assurance in design, development, production, installation and servicing'

JACOBSON, I., BOOCH, G. and RUMBAUGH, J.: 'The unified software development process' (Addison Wesley Publishing Inc, 1999)

KRUCHTEN, P.: 'The Rational Unified Process: an introduction' (Addison-Wesley Publishing Inc, 1999)

MAZZA, C., FAIRCLOUGH, J., MELTON, B., DE PABLO, D., SCHEFFER, A. and STEVENS, R.: 'Software engineering standards' (Prentice-Hall, 1994)

Other useful resources

The following is a short list of some very useful resources on the Internet. Although the list is brief, the addresses form an excellent starting point for looking for any sort of information on systems engineering, or the UML.

WWW.INCOSE.ORG (The official website for the International Council on Systems Engineering who have an excellent set of links to almost all aspects of systems engineering)

WWW.IEE.ORG (The official website for the Institution of Electrical Engineers, who provide may systems engineering resources including access to the Professional electronic Network for Systems Engineers (PeNSE))

WWW.OMG.ORG (The official website for the Object Management Group who have control over the UML standard. The actual standard may be downloaded from this site along with a lot of other UML information)